Rebecca Medda

STED microscopy as a tool in cell biology

Rebecca Medda

STED microscopy as a tool in cell biology

Investigation of age-related alterations occurring on the protein and the lipid level by using STED microscopy

Südwestdeutscher Verlag für Hochschulschriften

Impressum/Imprint (nur für Deutschland/ only for Germany)
Bibliografische Information der Deutschen Nationalbibliothek: Die Deutsche Nationalbibliothek verzeichnet diese Publikation in der Deutschen Nationalbibliografie; detaillierte bibliografische Daten sind im Internet über http://dnb.d-nb.de abrufbar.

Alle in diesem Buch genannten Marken und Produktnamen unterliegen warenzeichen-, marken- oder patentrechtlichem Schutz bzw. sind Warenzeichen oder eingetragene Warenzeichen der jeweiligen Inhaber. Die Wiedergabe von Marken, Produktnamen, Gebrauchsnamen, Handelsnamen, Warenbezeichnungen u.s.w. in diesem Werk berechtigt auch ohne besondere Kennzeichnung nicht zu der Annahme, dass solche Namen im Sinne der Warenzeichen- und Markenschutzgesetzgebung als frei zu betrachten wären und daher von jedermann benutzt werden dürften.

Verlag: Südwestdeutscher Verlag für Hochschulschriften Aktiengesellschaft & Co. KG
Dudweiler Landstr. 99, 66123 Saarbrücken, Deutschland
Telefon +49 681 37 20 271-1, Telefax +49 681 37 20 271-0
Email: info@svh-verlag.de
Zugl.: Heidelberg, Ruprecht-Karls Universität, Diss., 2009

Herstellung in Deutschland:
Schaltungsdienst Lange o.H.G., Berlin
Books on Demand GmbH, Norderstedt
Reha GmbH, Saarbrücken
Amazon Distribution GmbH, Leipzig
ISBN: 978-3-8381-1538-2

Imprint (only for USA, GB)
Bibliographic information published by the Deutsche Nationalbibliothek: The Deutsche Nationalbibliothek lists this publication in the Deutsche Nationalbibliografie; detailed bibliographic data are available in the Internet at http://dnb.d-nb.de.

Any brand names and product names mentioned in this book are subject to trademark, brand or patent protection and are trademarks or registered trademarks of their respective holders. The use of brand names, product names, common names, trade names, product descriptions etc. even without a particular marking in this works is in no way to be construed to mean that such names may be regarded as unrestricted in respect of trademark and brand protection legislation and could thus be used by anyone.

Publisher: Südwestdeutscher Verlag für Hochschulschriften Aktiengesellschaft & Co. KG
Dudweiler Landstr. 99, 66123 Saarbrücken, Germany
Phone +49 681 37 20 271-1, Fax +49 681 37 20 271-0
Email: info@svh-verlag.de

Printed in the U.S.A.
Printed in the U.K. by (see last page)
ISBN: 978-3-8381-1538-2

Copyright © 2010 by the author and Südwestdeutscher Verlag für Hochschulschriften Aktiengesellschaft & Co. KG and licensors
All rights reserved. Saarbrücken 2010

To my family

Contents

1 Introduction   **15**
    1.1 Fluorescence Microscopy in Cell Biology 15
         1.1.1 Fundamentals of Light Microscopy in Cell Biology 15
         1.1.2 Fluorescence Microscopy 16
         1.1.3 Specific Labeling 18
         1.1.4 Fluorescence Correlation Spectroscopy 19
    1.2 Fluorescence Nanoscopy 25
         1.2.1 STED Microscopy 26
    1.3 The Plasma Membrane 28
         1.3.1 The Phospholipids and Cholesterol 29
         1.3.2 The 'Raft' Controversy 32
    1.4 Proteins Interacting with Constituents of the Plasma Membrane 33
    1.5 The Human Neuroblastoma Cell Line SH-SY5Y 34
         1.5.1 Redifferentiation of SH-SY5Y 34
    1.6 The Cytoskeleton 35
         1.6.1 The Neuronal Cytoskeleton 36
         1.6.2 Organization of Neurofilaments 37
         1.6.3 Post-translational Modifications of Neurofilaments 39
         1.6.4 Role in Neurodegeneration 41

2 Motivation   **43**

3 Results   **45**
    3.1 Post-translational modifications of neurofilaments visualized by STED
        Microscopy 45

Contents

		3.1.1	Appearance of Intermediate Filament Proteins during Redifferentiation of SH-SY5Y	46
		3.1.2	Reciprocal Expression of Vimentin and Doublecortin	49
		3.1.3	Visualization of Post-Translational Modifications of Neurofilaments using STED Microscopy	50
		3.1.4	Hyperphosphorylation upon Glucose Deprivation	51
		3.1.5	The Role of p38 Mitogen Activated Protein Kinase in Glucose Deprived cells	55
		3.1.6	The Role of c-Jun N-terminal Kinase in Glucose Deprived cells .	57
	3.2	Lipid Diffusion in the Plasma Membrane	59
		3.2.1	Reduction of the Focal Volume by Combining STED with Fluorescence Correlation Spectroscopy	60
		3.2.2	Incorporation of Lipids into the Plasma Membrane	60
		3.2.3	The Influence of the Fluorescent Label on the Lipid Behavior . .	66
		3.2.4	Measurements of Lipid Dynamics in the Plasma Membrane . . .	72
		3.2.5	Influence of Cholesterol on the Diffusion of SM, GM1 and Cer . .	74
		3.2.6	Effectivity of Cholesterol Depletion	77
		3.2.7	Approximation of the Spatial Dimensions of Microdomains . . .	78
4	**Discussion and Outlook**			**81**
	4.1	Post-translational Modifications of Neurofilaments	81
		4.1.1	The Effects of Retinoids on Cancer Cells	81
		4.1.2	Phosphorylation of Neurofilaments	83
		4.1.3	Phosphatases in Neurodegeneration	83
		4.1.4	Responses to Metabolic Stress	85
		4.1.5	Viability of Deprived Cells	87
		4.1.6	Double Staining of Phosphorylated and O-GlcNAcylated Epitopes	87
		4.1.7	Investigation of the Post-Translational Modifications using STED Microscopy	88
	4.2	Incorporation of Labeled Lipids in the Plasma Membrane	90
		4.2.1	Fusion of Liposomes with the Plasma Membrane	90
		4.2.2	Plasma Membrane Sheet Generation	91
		4.2.3	Serum Albumins - Lipid Carriers in the Blood	91

	4.2.4	Proper Incorporation of the Labeled Lipids in the Plasma Membrane	92
	4.2.5	Diffusion of Phospholipids with Saturated and Unsaturated Fatty Acids	94
	4.2.6	Outer or Inner Leaflet?	94
	4.2.7	Cholesterol Depletion	95
	4.2.8	The Interactions of the Cytoskeleton with Constituents of the Plasma Membrane	96
	4.2.9	Temperature Dependence on Lipid Diffusion	97
	4.2.10	Potential Artifacts due to the High Intensities of the STED and Excitation Lasers	98
4.3	Specific Labeling in High Resolution Microscopy		100
	4.3.1	Alternative Affinity Markers	100
	4.3.2	*in vitro* Labeling of Recombinant Proteins	103
	4.3.3	N-V Centers: Extremely Stable Labels in High Resolution Microscopy	105

5 Materials and Methods — 109

5.1	Buffers and Solutions		109
5.2	Lipids		109
	5.2.1	Liposome generation	110
	5.2.2	Generation of Plasma Membrane Sheets	111
	5.2.3	Formation of the lipid-BSA complex	112
	5.2.4	Cholesterol depletion	115
5.3	Organisms and organism specific methods		116
	5.3.1	Cultivation of *Escherichia coli*	116
	5.3.2	Transformation of *E. coli*	117
	5.3.3	Isolation and purification of plasmid DNA from *E. coli*	117
	5.3.4	Mammalian cell lines	118
	5.3.5	Media and cultivation	119
	5.3.6	Re-differentiation of SH-SY5Y	120
	5.3.7	Test on mycoplasma	120
	5.3.8	Transfection of mammalian cells	121
5.4	Protein biochemistry		121

Contents

		5.4.1	Western analysis	122
		5.4.2	Kinase assays	125
	5.5	Specific labeling		127
		5.5.1	DNA dyes, TMA-DPH and mito trackers	127
		5.5.2	Dye Conjugation of proteins	129
		5.5.3	Immunofluorescence	131
	5.6	Light microscopy		134
		5.6.1	Wide field Fluorescence microscopy	134
		5.6.2	Confocal microscopy	134
		5.6.3	STED-FCS setup	135
		5.6.4	Supercontinuum STED	136

6 Abbreviations   **139**

Bibliography   **164**

7 Supplementary   **165**

8 List of Publications   **169**
  8.1 Publications Related to Thesis 169
  8.2 Publications in Cooperation (Selection) 170

9 Contribution to Conferences   **171**

10 Acknowledgment   **173**

Summary

High resolution light microscopy is gaining importance in the Life Sciences. This thesis reports on two applications of the STED (stimulated emission depletion) principle investigating alterations on the protein and lipid level occurring during aging.

One hallmark of age-related neurodegenerative diseases is the abnormal post-translationally occurring hyperphosphorylation of proteins. Here, the interplay of two main post-translational modifications of neurofilaments, phosphorylation and glycosylation, was investigated. The organization pattern of these modifications can not be resolved by conventional diffraction-limited light microscopy. However, with the about 10-times higher resolution of STED microscopy, the organization for both modifications was revealed. Long term glucose deprivation of redifferentiated neuroblastomas results in axonal swellings and cell death, which could not be rescued by refeeding glucose. These severe alterations could be observed by confocal resolution ($\sim$250 nm). Taking advantage of the higher resolution of the STED microscope (20-30 nm) allowed the detection of earlier alterations in the pattern of both modifications, where cell death could be prevented by the refeeding of glucose. Two kinases were found to be involved in the phosphorylation events occurring upon glucose deprivation in redifferentiated SH-SY5Y: the mitogen activated protein kinase p38 MAPK and to a lower extent also the stress activated protein kinase JNK.

An effect occurring during aging is the change in the plasma membrane's fluidity. This thesis reports on the investigation of the diffusion characteristics of different phospholipids depending on the endogenous cholesterol content on a nanoscale. Fluorescent mono- and diacyl-phospholipid analogs were successfully incorporated via BSA-complex formation into the plasma membrane of living cells. It was found that for proper incorporation a molar ratio of 1:1 (fatty acid chains:BSA molecules) had to be used. Analogs of the phosphoglycerolipids and the sphingolipids, respectively, were the central study in this thesis. It was shown that sphingolipids, in contrast to phosphoglycerolipids, were trapped on spatial scales below 30 nm, which could not be examined by confocal microscopy. Upon changing the endogenous cholesterol level, enzymatically with cholesterol oxidase and by host-guest complex formation using the oligosaccharide β-cyclodextrin, respectively, this hindered-diffusion could be reversibly abolished.

Zusammenfassung

Hochauflösende Mikroskopieverfahren gewinnen zunehmend an Bedeutung in den Lebenswissenschaften. In dieser Arbeit werden zwei Anwendungen des STED (Stimulated Emission Depletion) Prinzips vorgestellt, die sich mit der Untersuchung von altersbedingten Veränderungen auf der Protein- und Lipidebene befassen.

Ein Kennzeichen altersbedingter neurodegenerativer Erkrankungen ist eine abnorme post-translationelle Hyperphosphorylierung von Proteinen. In dieser Arbeit wurde das Wechselspiel zwischen den zwei häufigsten post-translationellen Modifizierungen von Neurofilamenten, Phosphorylierung und Glykosylierung, untersucht. Die Art der Organisation dieser Modifizierungen kann nicht mit herkömmlichen beugungsbegrenzten Mikroskopieverfahren aufgelöst werden. Die allerdings 10-fach höhere Auflösung der STED Mikroskopie ermöglichte die Darstellung beider Modifizierungen. Langanhaltender Glukosemangel bei redifferenzierten Neuroblastoma führte zu axonalen Schwellungen und schliesslich zum Zelltod, der durch die Zugabe von Glukose nicht verhindert werden konnte. Diese schwerwiegenden Veränderungen konnten bereits mit konfokaler Auflösung ($\sim$250 nm) beobachtet werden. Unter Ausnutzung der höheren Auflösung der STED Mikroskopie (20-30 nm), konnten Veränderungen in dem Muster beider Modifizierungen schon früher festgestellt werden. Zu diesem Zeitpunkt konnte die Zugabe von Glukose den Zelltod noch aufhalten. Zwei Kinasen zeigten einen Aktivitätsanstieg ausgelöst durch Glukosemangel in redifferenzierten SH-SY5Y: die Mitogen aktivierte Proteinkinase p38 MAPK und in geringerem Ausmaß die Stress aktivierte Proteinkinase JNK.

Ein weiterer Effekt, der altersbedingt auftritt, ist die Veränderung in der Fluidität der Plasmamembran. In dieser Arbeit wurden die Diffusionseigenschaften verschiedener Phospholipide in Abhängigkeit des endogenen Cholesterolspiegels auf kleinster Skala untersucht. Fluoreszierende Mono- und Diacylphospholipidanaloga wurden stabil durch BSA-Komplexbildung in die Plasmamembran lebender Zellen eingegliedert. Für eine stabile Eingliederung musste ein molares Verhältnis von 1:1 zwischen den Fettsäureketten und den BSA Molekülen herrschen. Phosphoglycerolipid- und Sphingolipidanaloga standen im Mittelpunkt der Untersuchungen. Es konnte gezeigt werden, dass Sphingolipide im Gegensatz zu den Phosphoglycerolipiden auf Skalen kleiner 30 nm gehinderte Diffusion (trapping) aufwiesen. Dies konnte nicht mit konfokaler Mikroskopie beobachted werden. Nach Veränderung des endogenen Cholesterolspiegels, entweder enzymatisch durch Cholesterol Oxidase oder durch Wirt-Gast Wechselwirkungen von β-Cyclodextrin mit Cholesterol, konnte diese gehinderte Diffusion aufgehoben werden.

1 Introduction

1.1 Fluorescence Microscopy in Cell Biology

In Life Sciences fluorescence microscopy is nowadays a commonly used technique gaining more and more in importance. Its impact is also reflected in the number of relevant publications. A search in *Web of Science* for the term 'microscopy' leads to more than 87,000 results for the year 2008. And by having a closer look at the distribution of light and electron microscopy throughout renowned journals in the field of the life sciences, light microscopy covers about approximately 80%.

1.1.1 Fundamentals of Light Microscopy in Cell Biology

The extent of a typical animal cell is about 10-20 μm, which lies far beyond of what is visible for the human eye. A great demand on making cells visible led to an evolution of enlarging scopes based on optics. The first described microscope, consisting of a converging objective and a diverging eye piece was built by the two spectacles makers Zacharias and Hans Janssen around 1590. Shortly after, in 1609, Galileo Galilei worked out the principles of lenses and further enhanced it by introducing a focusing lens.

As one of the first applications, Robert Hooke observed cells in cork tissue by using a compact microscope in 1665 (Hooke, 1665). It was then, when he coined the expression *cell* for they reminded him of the cells in a monastery. Shortly after, in 1674, Antonie van Leeuwenhoek was able to see single algae cells with his self built, improved microscope. With the discovery of Schleiden and Schwann in 1838, that plants and animals consist of several individual cells, cell biology was born.

It was Ernst Abbe, who finally developed a physical description of the theory of microscopy in 1873, stating that the minimal distance d, at which two points can be still separated from each other is essentially limited by the wavelength λ of the employed light

Introduction

due to diffraction (Abbe, 1873). This law is known as the Abbe formula:

$$d \approx \frac{\lambda}{2n \, sin\alpha}$$

where n is the refractive index of the embedding medium and α the semi-aperture angle of the focused light. With NA=$n \, sin\alpha$, the minimal size of the focus is governed by the wavelength and the numerical aperture NA of the system. Together with Otto Schott, Ernst Abbe finally developed improved the manufacturing process so that their microscopes could operate only limited by the diffraction law, stated above.

1.1.2 Fluorescence Microscopy

Many questions in cell biology demand a visualization of certain cellular structures or proteins. Because animal cells are usually colorless and translucent, there was a need for stains which fulfill two main requests: it had to provide sufficient contrast and it had to be specific for a distinct structure.

The first criterion is fulfilled by the utilization of the physical phenomenon of fluorescence, which is restricted to the fluorescent molecules only, resulting in a high contrast. The second criterion, the specific labeling, is reviewed in the next subsection. The principles of fluorescence were observed and described by Sir George Gabriel Stokes in his scientific paper 'On the refrangibility of light' in 1852 (Stokes, 1852).

The different electronic states and transitions, which participate in fluorescence were described by Jablonski (see figure 1.1). Fluorescent molecules are excited by photons of a certain wavelength from the singlet ground state S_0 into the first excited singlet state S_1. Within nanoseconds (equivalent to the fluorescent life time) they spontaneously relax back to the S_0 state by emitting photons with a longer wavelength. The difference of excitation and emission wavelength is the cause of the above mentioned high contrast and is called Stokes shift. By using appropriate filters the emission light can be told apart from the excitation light.

Additionally, a transition from S_1 to the first triplet state T_1 is also possible though to a much lower extent. Since this transition is spin forbidden, it happens on a longer time scale (within microseconds). The triplet state is the molecules long living dark state and is depopulated within microseconds to milliseconds (since this transition is also spin forbidden) to the ground state S_0.

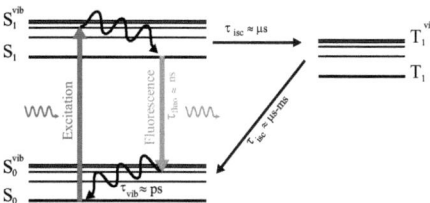

Figure 1.1: Jablonski diagram of a fluorescent molecule. The electronic singlet (S) and triplet (T) states of the first order and additional vibrational levels are shown. Molecules are excited by absorption of a photon (green wiggly line) from the electronic ground state S_0 to a vibrational level of the first excited electronic state S_1. By quick internal relaxation ($\sim 10^{-12}$ s) they reach the vibrational ground state of S_1. From there, spontaneous fluorescence may occur within 10^{-9} s by emitting a photon with a longer wavelength (yellow wiggly line) due to the Stokes shift. The electronic triplet state T_1 is populated via intersystem crossing (isc) from the S_1. For this transition is spin forbidden, it takes place in the microsecond range. The molecules spontaneously relax from their long living dark state ($\sim 10^{-6}$ s -10^{-3} s) back to the ground state radiationless by thermal relaxation (in aqueous solution).

In the application of this phenomenon to light microscopy another important step had to be taken: the development of a fluorescence microscope. The principle of it was already recognized by August Köhler in 1904, but for opening it to the scientific community, the need for specifically labeling structures in cells, was still an open question. Finally, in the 1940s, Albert H. Coons and Melvin H. Kaplan finally succeeded in conjugating fluorescent dyes to antibodies, therefore breaking it to the field of life sciences (Coons and Kaplan, 1950). Since then, several different strategies of using fluorescent dyes as specific labels in optical microscopy have been developed.

The commonly used wide field based microscopy, however, had a bad z discrimination (along the optical axis). Finally, in the 1960s, Marvin Minsky patented the confocal microscope (Minsky, 1961). An introduction of point illumination and a pinhole in front of the detector, provided z-discrimination and thus increased the image quality. By introduction of this pinhole, optical sectioning becomes possible, allowing a three dimensional reconstruction of the sample. Not until the end of the 1980s, after the subsequent development of powerful stable light sources (lasers) and sensitive detectors (photomultiplier tubes, PMTs) fluorescence confocal microscopy became a widely used technique (Amos et al., 1987). Before the introduction of this technique, in near-field optics one was restricted surface-near specimens. The use of far-field optics, such as the above described,

Introduction

allows an investigation of structures which are not exclusively on the surface of the sample.

1.1.3 Specific Labeling

In light microscopy, unlike in biochemical methods or electron microscopy, the subcellular localization and distribution of the molecules of interest can be studied in the intact cells due to the non-invasive nature of the used light. The investigation of cellular and subcellular structures on the basis of image-guided methods in optical microscopy (imaging) requires techniques that allow for a specific detection and recording.

For fixed cells or tissues the most commonly used method is the direct or indirect immunofluorescence (see figure 1.2) to address the fluorescent label to the desired structure. This enables specific labeling of the accessible epitopes on the endogenous structures. The two step incubation used in indirect immunofluorescence leads to amplification of the fluorescence signal, since several secondary antibodies can bind to one primary. Furthermore, only a small set of commonly used secondary antibodies (e.g. anti-mouse, anti-rabbit and anti-goat) has to be conjugated instead of every single primary antibody. The direct method, however, has the advantage to use several different dye-conjugated primary antibodies in multicolor applications, although raised in the same species.

However, the use of affinity markers like antibodies is restricted to mostly fixed cells, because of their tendency to form large complexes with their targets. Thus, life cell markers had to be introduced.

Genetically encoded fluorescent fusion proteins, whose DNA sequence is fused to the one of the target protein, can be designed and introduced into the cells via plasmids. The discovery, cloning and improvement of the green fluorescent protein (GFP) was awarded with the Nobel Prize in chemistry to Osamu Shimomura, Martin Calfie and Roger Tsien in 2008. Nowadays, fluorescent proteins (FPs) are available covering the whole visible spectrum. In most applications, however, the introduction of FPs implicates an uncontrollable amount of the tagged protein in addition to the endogenous protein level (if not under the control of the endogenous promoter), resulting in over-expression. Furthermore, the influence of the fluorescent protein fused to the protein of interest regarding its localization or distribution is not fully understood and can not be excluded. DNA specific dyes like DAPI, strongly enhance their fluorescence upon binding to double stranded DNA,

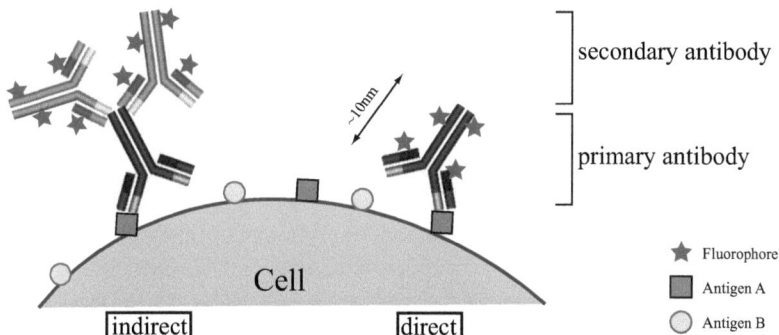

Figure 1.2: Indirect and direct immunofluorescence. The indirect detection method uses two types of antibodies, primary and secondary antibodies, whereas the direct method only uses primary antibodies. The primary antibody is only specific for antigen A (blue). A subsequent, dye-conjugated secondary antibody specifically binds the organism-specific c-terminal region of the primary antibody. In direct immunofluorescence the primary antibody is conjugated with the fluorophore forthright.

therefore providing an excellent signal-to-noise ratio. For life cell imaging, some fluorophores, e.g. Mitotracker or Lysotracker, are targeted into mitochondria or lysosomes, respectively. Whereas other dyes are sensitive for the membrane potential, such as JC-1, which is especially sensitive for the mitochondrial membrane potential. These dyes can therefore be used as reporters to give information about the state of the cell's viability.

Besides the imaging aspect, fluorescent molecules provide a whole variety of additional readouts, such as the fluorescence lifetime, the fluorescence anisotropy and the fluorescence intensity (Lakowicz, 1999; Eggeling et al., 2001). The next section gives a brief introduction of the spectroscopic aspects of fluorescence microscopy in general and fluorescence correlation spectroscopy in particular.

1.1.4 Fluorescence Correlation Spectroscopy

The theoretical description of fluorescence has besides imaging also led to the development of many spectroscopic techniques with high temporal resolution (Orrit and Bernard, 1990; Shera et al., 1990; Moerner and Fromm, 2003; Zander et al., 2002). The investigation of dynamical processes in the living cell requires the application of techniques, which provide a high spatial and temporal resolution. Fluorescence correlation spec-

Introduction

troscopy (FCS) can be used to determine local concentrations, mobility coefficients or characteristic rate constants of fluorescently labeled molecules by analyzing their fluorescence intensity fluctuations. This combines the advantages of fluorescence microscopy (the rather high spatial resolution and the non-invasiveness) with the high temporal resolution obtained by correlation spectroscopy analysis.

Another, non-correlative approach is the also widely used technique fluorescence recovery after photobleaching (FRAP). In FRAP a small area or volume is rapidly photobleached and the recovery of fluorescence intensity is recorded while bleached fluorophores diffuse out and unbleached fluorophores diffuse in that area or volume, respectively (Axelrod et al., 1976). The diffusion constant D can be determined from diffusion times, however, its ability to distinguish between different diffusion constants is particular problematic for two-dimensional diffusion (e.g. in the plasma membrane). Another important issue to be checked is the possible photodamage, induced by the high level of bleaching light.

However, unlike FCS it is not concentration limited and also unlike conventional FCS it can measure and distinguish mobile fractions of fluorophores from immobile ones. It also provides a reliable tool for verifying proper incorporation of lateral diffusing fluorophores in a two-dimensional system.

In the 1970s, the concept of FCS was introduced by the group of Watt W. Webb (Elson and Magde, 1974; Madge et al., 1974). Unlike the sample preparation applied for fluorescence imaging, the labeling density in FCS experiments needs to be very small, since the overall fluorescence fluctuations of many fluorophores are small or not detectable compared to the total signal. A too sparsely labeled sample, on the other side, leads to a greatly increased measurement time. Further requirements are a low background and a good signal from the fluorophore. The background stemming from outer focal planes could successfully be suppressed in axial direction by using confocal detection (Rigler and Widengren, 1990; Rigler et al., 1993). Thus the detection volume could be decreased to less than a femtoliter (10^{-15} l), allowing fluorophore concentrations in the nanomolar to micromolar range ($10^{-9} - 10^{-6}$ M).

In FCS experiments, the sample is illuminated and the intensity traces from single fluorophores are recorded over time. These fluctuations in fluorescence intensity are analyzed using the temporal autocorrelation. These fluctuations are due to inter- or intramolecular reactions, by rearrangements of the microenvironment of the fluorophore or due to diffusion. The term 'spectroscopy' in FCS was actually coined by the temporal autocorrelation, for it is a genuine form of spectrum (a time-spectrum) generated from the power spectrum via inverse Fourier transformation.

Theory of FCS

Since the theory of FCS is described in details in many reviews (Madge, 1976; Rigler and Elson, 2001; Schwille et al., 1997), only an overview about the concept is given here. Two models, the standard model and the extended model, describing single molecule diffusion and hindered diffusion, respectively, are introduced in the following.

Standard model: Single molecule diffusion. Assuming constant excitation power, the temporal fluctuations in FCS are defined as the deviations of the temporal average of the fluorescence signal of fluorescent molecules diffusing through the focus $\delta F(t) = F(t) - \langle F(t) \rangle$, where $\langle F(t) \rangle$ describes the time-averaged part and $F(t)$ the fluorescence signal. The total fluorescence signal stems from the whole detection volume V:

$$F(t) = \sigma \cdot q \cdot \kappa \cdot \int_V I_{exc}(r) \cdot h_{det}(r) \cdot C(r,t) d^3r \qquad (1.1)$$

with the constants κ describing the overall detection efficiency of the microscope, σ describing the molecular absorption cross section and q describing the quantum yield of the fluorophore. $C(r,t)$ gives the spatial temporal distribution of the fluorophore concentration at the position $r = (x, y, z)$, whereas $I_{exc}(r)$ and $h_{det}(r)$ denote the spatial distribution of the excitation spot (PSF, point spread function) and the detection efficiency of the optical setup. They can be combined to the dimensionless brightness function $B(r) = I_{exc}(r)/I_0 \cdot h_{det}(r)$, with I_0 describing the maximum amplitude of the excitation intensity. In the confocal case, $B(r)$ can be well approximated by a three dimensional Gaussian function.

$$B(r(x,y,z)) = exp(-2[x^2 + y^2]/r_0^2)exp(-2z^2/z_0^2), \qquad (1.2)$$

with r_0 and z_0 depicting the lateral and axial coordinates, respectively, where the intensity has dropped to e^{-2} from its initial value.

The fluctuations in the fluorescence signal stem either from changes in the fluorophore

Introduction

concentration $\delta C(r,t) = C(r,t) - \langle C \rangle$, from changes in the absorption cross section $\delta \sigma$ or are due to changes in the quantum yield, they can be described by

$$\delta F(t) = \delta \sigma \cdot \delta q \cdot \kappa \cdot I_0 \cdot \int_V B(r) \cdot \delta C(r,t) d^3r. \tag{1.3}$$

The parameters κ, σ and q can be combined with the maximum excitation intensity amplitude I_0 to give a measure for the signal-to-noise ratio. This parameter then denotes the photon count rate per detected molecule per second. With $\eta = I_0 \cdot \sigma \cdot q$ equation (1.3) can be written as:

$$\delta F(t) = \int_V B(r) \cdot \delta(\eta \cdot C(r,t)) d^3r \tag{1.4}$$

To reveal characteristic time constants of the diffusing molecules, the fluctuations in fluorescence signal can be quantified by calculating the normalized second order autocorrelation of the acquired fluorescence intensity.

$$G(\tau) = \frac{\langle F(t) F(t+\tau) \rangle^2}{\langle F(t) \rangle} = 1 + \frac{\langle \delta F(t) \cdot \delta F(t+\tau) \rangle}{\langle F(t) \rangle^2} = 1 + \frac{\langle \delta F(0) \cdot \delta F(\tau) \rangle}{\langle F \rangle^2}, \tag{1.5}$$

with correlation time τ and the triangular brackets indicating averaging over the measurement time. Upon insertion of equation (1.4) equation (1.5), the autocorrelation function can be written as:

$$G(\tau) = 1 + \frac{\int \int B(r) B(r') \langle \delta(\eta \cdot C(r,t)) \cdot \delta(\eta \cdot C(r',t+\tau)) \rangle d^3r d^3r'}{\int B(r) (\langle \delta(\eta \cdot C(r,t)) \rangle d^3r)^2} \tag{1.6}$$

The assumption is made, that the photophysical properties of the fluorescent molecules are not changed during the data acquisition, leading to $\delta \eta = 0$. The measured fluorescence fluctuations are then only due to diffusion through the focal volume. This simplifies the equation above to:

$$G(\tau) = 1 + \frac{\int \int B(r) B(r') \langle \delta C(r,0) \cdot \delta C(r',\tau) \rangle d^3r d^3r'}{(\langle C \rangle \int B(r) d^3r)^2} \tag{1.7}$$

With fluorophores freely diffusing in three dimensions with the diffusion coefficient D,

1.1.4 Fluorescence Correlation Spectroscopy

the concentration autocorrelation term $\langle \delta C(r,0) \cdot \delta C(r',\tau) \rangle$ is given by

$$\langle \delta C(r,0) \cdot \delta C(r',\tau) \rangle = \langle C \rangle \, (4\pi D\tau)^{\frac{3}{2}} \cdot e^{-\frac{(r-r')^2}{4D\tau}}. \tag{1.8}$$

A more commonly used form of this equation can be obtained considering the following: firstly, the relationship between the diffusion time τ_D and the diffusion coefficient D is given by

$$\tau_D = \frac{r_0^2}{4D}, \tag{1.9}$$

and secondly, the effective volume (which is assumed to be Gaussian-shaped) is calculated by $V_{eff} = \pi^{\frac{3}{2}} \cdot r_0^2 z_0$ Combining finally these considerations with equation (1.8) leads directly to the normalized autocorrelation function (assuming a Gaussian-shaped function $B(r)$):

$$G(\tau) = \frac{1}{V_{eff} \langle C \rangle} \cdot \frac{1}{1 + \frac{\tau}{\tau_D}} \cdot \frac{1}{\sqrt{1 + (\frac{r_0}{z_0})^2 \cdot \frac{\tau}{\tau_D}}} \tag{1.10}$$

The amplitude $G(0) = \frac{1}{V_{eff} \langle C \rangle}$ of this autocorrelation function is inversely proportional to the average number of fluorescent particles $N = V_{eff} \langle C \rangle$ in the total volume. Therefore, if the measurement volume V is known, the average local molecule concentration can be calculated from the amplitude $G(0)$ by $\langle C \rangle = \frac{1}{V \cdot G(0)}$.

For fluorescent molecules which do not diffuse in a three dimensional volume but are restricted to a two dimensional system, the autocorrelation curve can be further simplified to

$$G_{2D}(\tau) = \frac{1}{V_{eff} \langle C \rangle} \cdot \frac{1}{(1 + \tau/\tau_D)}. \tag{1.11}$$

From this the diffusion coefficient can be easily determined from the characteristic decay time of the autocorrelation curve from τ_D.

Extended model: Hindered diffusion. In quasi two dimensional lipid membranes and inside living cells, free diffusion cannot be assumed, because the molecule's diffusion is restricted or is changed locally, due to compartimentalization or specific or nonspecific interactions. The diffusion described in the standard model was assumed to be free, meaning that the mean square displacement of the of the molecules increases linearly with the time $\langle r^2 \rangle \sim t$. For molecules which are hindered in their diffusion due to obstacles, barri-

Introduction

ers or trapping events, the mean square displacement is not anymore directly proportional to time but can be described with the relationship $\langle r^2 \rangle \propto t^\alpha$, with $\alpha < 1$ (Saxton, 1994; Schwille et al., 1999; Wachsmuth et al., 2000). This effects the distribution of focal transit times, which could be described by one diffusion constant D in the free diffusion model. There are in principle two possibilities to overcome this: One method assumes not only one but several classes of diffusing molecules, each representing a fraction A_i of the total number of molecules with different diffusion constants D_i and diffusion times τ_D^i. In most cases it is sufficient to assume two classes of molecules. This two component diffusion model can be described according to (Schwille et al., 1999):

$$G_{3D}(\tau) = \frac{1}{V_{eff}\langle C \rangle}[A_1 \cdot \frac{1}{1+\tau/\tau'_{xy}} \cdot \frac{1}{\sqrt{1+\tau/\tau'_z}} + A_2 \cdot \frac{1}{1+\tau/\tau''_{xy}} \cdot \frac{1}{\sqrt{1+\tau/\tau''_z}}], \quad (1.12)$$

with $0 < A_1 < 1$ describing the fraction of freely diffusing molecules and $0 < A_2 < 1$ giving the fraction of molecules that are hindered or trapped ($A_1 + A_2 = 1$) and with τ' and τ'' defining the corresponding focal transit times.

The other possibility is the introduction of an anomaly factor $\alpha < 1$ into the autocorrelation function, where smaller values indicate a higher anomaly (Schwille et al., 1999), leading to

$$G_{3D}(\tau) = \frac{1}{V_{eff}\langle C \rangle} \cdot \frac{1}{1+(\frac{\tau}{\tau_D})^\alpha} \cdot \frac{1}{\sqrt{1+[\frac{r_0}{z_0}^2 \cdot \frac{\tau}{\tau_D}]^\alpha}}. \quad (1.13)$$

Both models can be used for describing hindered or trapped diffusion, where the two species model is more descriptive and the anomaly model is more rigorously but its results less intuitive. For all the previous considerations all fluctuations in the fluorescence were assumed being solely due to diffusion of the molecule. To calculate the autocorrelation curve for molecules, whose brightnesses are also affected by inter- or intramolecular reactions, two additional terms have to be introduced to account for these changes, leading to a product of different autocorrelation function components according to each kinetic reaction $G_{kinetics}(\tau)$.

$$G(total) = G_{Diff(\tau)} \cdot \prod G_{kinetics}(\tau). \quad (1.14)$$

The most common reason for alterations in the fluorescence brightness is the population of the molecules' dark states such as the triplet state. In the following equation the autocorrelation function takes the triplet state T and its correlation time τ_T into account (Widengren et al., 1994):

$$G(total) = G_{Diff}(\tau) \cdot [\frac{1-T}{T} \cdot \exp(-\frac{\tau}{\tau_T})]. \tag{1.15}$$

But, however, still two limitations are pivotal in both fluorescence techniques, imaging and spectroscopy: the photobleaching of the dye and the finite resolution of visible light due to diffraction. The usage of photostable dyes and triplet quenching agents or oxygen scavenger systems circumvent the problem of photobleaching. The diffraction barrier of light microscopy, which Abbe depicted over 100 years ago, can be overcome by using electron microscopy (Knoll and Ruska, 1932) or near-field techniques, such as scanning near-field optical microscopy (SNOM) (Synge, 1928; Durig et al., 1986), super-lenses (Pendry, 2000) and probing on nanostructures (Levene et al., 2003; Wenger et al., 2007). However, the major drawback of these techniques is the very low working distance (near-field techniques) demanding exclusively the observation of objects close to the surface or the harsh sample treatment (electron microscopy).

1.2 Fluorescence Nanoscopy

Although near-field techniques or electron microscopy, as depicted above, have already enabled a higher resolution, these were not applicable in the life sciences, where mostly far-field optical light microscopy techniques are applied. The lateral resolution in a conventional scanning far-field fluorescence microscope with a high numerical aperture (NA) lens (e.g. an oil objective lens with $NA = 1.42$) using an excitation laser light of e.g. $\lambda_{exc} = 633$ nm has a theoretical resolution limit of $d \approx 223$ nm according to $d \approx \frac{\lambda_{exc}}{2n \sin \alpha}$, with $n \sin \alpha = NA$.

Several techniques have been developed in the last 15 years for breaking the diffraction barrier in optical far-field microscopy which are reviewed in (Hell, 2007). They all share a common principle: the (reversible) switching of an optical property of a molecule between two distinguishable states A and B. These optical properties can be the fluorescence itself (bright and dark states), the absorption (absorbing and non-absorbing), the orientation of the molecule (perpendicular and parallel) and others. In the RESOLFT (REversible Saturated OpticaL Fluorescence Transitions) concept fluorescent molecules un-

Introduction

dergo transitions between two states, e.g. the ground state S_0 and the first singlet excited state S_1. The triplet state defines the molecule's dark state and is populated by intersystem crossing from the singlet state, however with a decreased probability, since this transition is spin forbidden. The RESOLFT techniques can be classified into the targeted readout methods, with STED (STimulated Emission Depletion) microscopy (Hell and Wichmann, 1994) as the most successfully realized example and GSD (Ground State Depletion) (Hell and Kroug, 1995; Bretschneider et al., 2007), and into stochastic methods, such as PALM (PhotoActivation Localization Microscopy), STORM (STochastic Optical Reconstruction Microscopy) and FPALM (Fluorescence PhotoActivation Localization Microscopy) (Betzig et al., 2006; Rust et al., 2006; Hess et al., 2006). The following paragraphs highlight the principles of STED microscopy in detail.

1.2.1 STED Microscopy

In 1994 the concept of STED microscopy have been proposed in theory (Hell and Wichmann, 1994) and were demonstrated experimentally in 1999 (Klar and Hell, 1999). The two optical distinguishable states A and B, demanded in RESOLFT, are reflected by the first excited state S_1 (bright) and by the ground state S_0 (dark) of the fluorophore. In STED, the size of the effective excitation focus is reduced by switching off the fluorescence of the fluorophores in the outer regions of the focus. This switching is light driven. As described in figure 1.3 **A**, the transition from S_0 to S_1^{vib} is induced upon absorption of a photon of a distinct wavelength. Fast internal relaxation to the vibrational ground state S_1^0 takes place within picoseconds. In the case of fluorescence, the transition to the electronic ground state S_0 occurs spontaneously within nanoseconds. On the other hand, under STED conditions, the inhibition of fluorescence by depopulating the S_1 state is driven by stimulated emission: upon incidence of a photon, which matches the energy gap between S_1 and S_0^{vib}, the molecule relaxes to S_0^{vib} by emitting a second photon with exactly the same properties (see figure 1.3). Thereby, the molecule is de-excited before spontaneous fluorescence occurs. To apply this phenomenon of stimulated emission in a microscopy scheme, the de-excitation of the fluorophore has to be spatially confined to the outer regions of the excitation focus. Therefore the STED focus needs a center with zero intensity and high intensities in the periphery, resulting in a doughnut-shaped PSF (point spread function). In the experimental setup such a focus can be realized by intro-

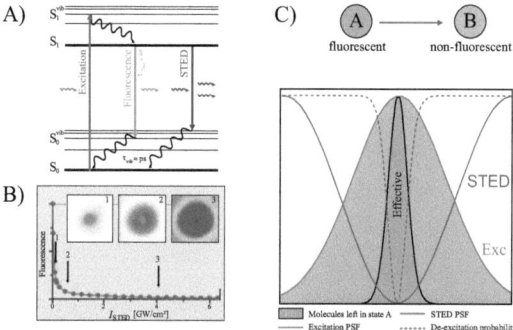

Figure 1.3: Breaking the diffraction barrier by STED microscopy. In **A** the electronic states involved in stimulated emission are shown. A fluorophore is excited upon absorption of a photon from the ground state S_0 into the excited state S_1. By spontaneous fluorescence, occurring within nanoseconds, it relaxes back into S_0. Another possibility to return to S_0 is by stimulated emission: a photon of a distinct wavelength (red wiggly line), whose energy matches the gap between the actual state and a state with lower energy (e.g. the highest vibrational state of the electronic ground state) induces the transition to the ground state by emitting a second photon of the same wavelength, thus inhibiting fluorescence. In **B** a typical fluorescence depletion curve and three accordant overlays of the excitation PSF (green) and the STED-PSF (red) are shown. Upon increase of the STED intensity I the effective excitation spot decreases. The line profiles shown in **C** depict the spatial regions where the fluorophore stays in A (orange area). Non-fluorescent fluorophores are shown in gray while the de-excitation probability is shown in blue. The transition probability from A (bright state) to B (dark state) is proportional to the applied STED intensity $I \gg I_{sat}$. This image was slightly modified after (Hell et al., 2004)

ducing an appropriate phase plate in the STED beam (Keller et al., 2007). The excitation focus is overlayed with the STED focus in a way that the central zero of the STED PSF matches the excitation PSF. This results in the depletion of fluorescence in the periphery, while from the center spot fluorescence still can occur. The higher the applied STED laser intensity I, the more efficient is the de-excitation of the fluorophores close to the center, thus leading to a theoretically infinitely small effective excitation spot (see figure 1.3 **B**).

Abbe's law can therefore be modified by a term taking the saturation of the STED intensity into account:

$$d \approx \frac{\lambda}{2n \sin\alpha \cdot \sqrt{1+\zeta}} \quad (1.16)$$

with $\zeta = I/I_{sat}$. The intensity of the STED beam, where the probability for the molecule is 1/2 to be in state A is called saturation intensity I_{sat} and is a dye specific constant. Upon increase of I the equilibrium is shifted towards the transition to the non-fluorescence state

B and the area where the dye is still in state A is in theory squeezed down to zero. In figure 1.3 the dependency of the applied STED intensity on the effective excitation spot is shown.

This reduced effective excitation spot is then scanned over the sample, leading to subdiffraction, resolved images. Furthermore, this technique can also be applied in spectroscopic techniques like FCS.

1.3 The Plasma Membrane

The plasma membrane is crucial for the life of the cell. It creates a diffusion barrier between the cell itself and its environment, for it is relatively impermeable to water-soluble molecules. However, it also mediates extracellular signals via receptors and is selectively permeable for certain molecules due to transmembranous protein channels or transporters. Besides the plasma membrane, in eucaryotic cells also intracellular compartments, like the golgi apparatus, the endoplasmic reticulum, mitochondria and other membrane-enclosed organelles maintain their difference to the cytosol by a lipid membrane. Although differing in their function, all biological membranes share a common assembly: lipids and proteins are connected via noncovalent interactions. The lipid molecules are arranged as a $\sim 5\,\mathrm{nm}$ thick double layer, the lipid bilayer, behaving like a two-dimensional fluid. Various classes of membrane proteins exist which mediate the interaction with the environment via receptors, active transport over the membrane and which also serve as structural anchor points for cytoskeletal elements.

All lipids are amphipathic, meaning, that they possess a hydrophobic and a hydrophilic part. In eucaryotic cells the phospholipids and sterols are the most abundant membrane lipids, consisting of a hydrophilic polar head group facing the aqueous environments (cytosol and extracellular surrounding) and a hydrophobic part (forming the actual diffusion barrier). In phospholipids, the hydrophobic domains are typically composed of normally two hydrophobic hydrocarbon tails (fatty acids consisting of 14 to 24 carbon atoms, where one is usually saturated and the other one unsaturated), whereas the hydrophobicity in cholesterol is due to the platelike steroid rings. The difference in length and saturation of the fatty acids determine the phospholipids' properties of diffusion and defines the fluidity by the packing ability against one another. The fluidity of the membrane is also dependent on the amount of cholesterol, which stiffens the membrane due to the interaction of its rigid steroid rings with the fatty acids of the phospholipids closest to their heads. On the other hand, it also prevents the fatty acid chains from coming to close together and crystallize.

Phospholipids very rarely migrate between the two leaflets (monolayers) of the plasma membrane, also called flip-flop, but exchange their place quite frequently with their neighbors within the leaflet, resulting in an asymmetry between the two leaflets (Devaux, 1993). However, since phospholipids are mainly made in the cytosolic ER membrane, a special class of membrane-bound enzymes, the phospholipid translocators, mediate a rapid flip-flop into the other leaflet. The outer leaflet facing the extracellular fluid consists predominantly of phosphatidyl choline (PC), sphingomyelin (SM), and glycolipids, like gangliosides (e.g. GM1) whereas the inner leaflet contains mostly phosphatidyl ethanolamine (PE), phosphatidyl serine (PS), and to a lower extent also phosphatidyl inositol (PI). This asymmetry of the outer and the inner leaflet is functional highly important. For instance, phosphatidyl inositol is concentrated in the cytosolic leaflet, playing an important role in signaling. Cholesterol, on the other hand, is distributed in both leaflets.

1.3.1 The Phospholipids and Cholesterol

The main constituents of the plasma membranes of eucaryotic cells is mainly constituted of phospholipids, cholesterol and membrane proteins. This subsection is focused on the lipids only and not going into membrane proteins. The different phospholipid species can be sorted into three different classes, introduced here in brief: the phosphoglycerolipids, the sphingolipids and the glycolipids.

Phosphoglycerolipids

In phosphoglycerolipids the hydroxyl groups at C-1 and C-2 of glycerol are esterified to two fatty acids, whereas the C-3 hydroxyl group is esterified to phosphoric acid, resulting in the simplest phosphoglycerolipid, phosphatidate. The major phosphoglycerolipids are bound to an additional alcohol via an ester bond between the phosphate group and the hydroxy group of the alcohol. The most abundant members are phosphatidyl choline (PC), phosphatidyl ethanolamine (PE) and phosphatidyl serine (PS), whereas the inositol phospholipids such as phosphatidyl inositol (PI), are represented at lower concentrations.

Introduction

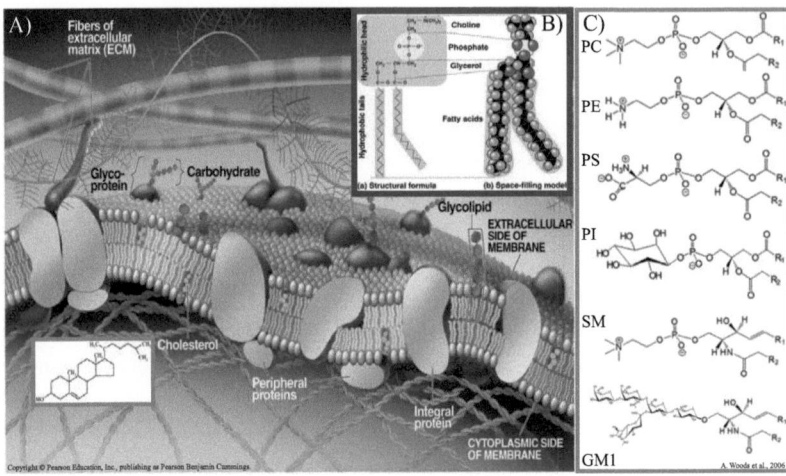

Figure 1.4: Lipid components of the plasma membrane. In **A** a sketch of a mammalian plasma membrane is shown including phospholipids, glycolipids, cholesterol, (trans-)membrane proteins and glycoproteins. In **B** the basic structure of a phospholipid, here phosphatidyl choline with one saturated and one mono-unsaturated fatty acid, is depicted. In **C** the structures of phosphoglycerolipids, sphingolipids, glycolipids, and cholesterol are shown. Phosphoglycerolipids contain two fatty acids esterified to glycerol, whereas the third carbon of glycerol is joined to a phosphate group (forming phosphatidic acid). This phosphate group in turn can be joined to another polar molecule (forming phosphatidyl ethanolamine, phosphatidyl choline, phosphatidyl serine, or phosphatidyl inositol). In sphingomyelin, the backbone is built by the complex alcohol sphingosine linked to a fatty acid over an amide bond (ceramide), whereas the hydrophilic head domain is formed by an ester bond from the primary hydroxyl group of sphingosine to phosphoryl choline.

As shown in figure 1.3.1 **C** they differ mainly in their head group structure, regarding size, shape and charge.

Sphingolipids

The sphingolipids, on the other hand have a sphingosine backbone, a more complex alcohol, that contains a long, unsaturated hydrocarbon chain. In the case of sphingomyelin (SM) the amino group of the sphingosine is linked to a fatty acid by an amide bond (ceramide), whereas the primary hydroxyl group of sphingosine is esterified to phosphoryl choline. For other ceramide backbone phosphosphingolipids, the head group can also be a phosphoryl ethanolamine. The function of sphingolipids still remains unclear, but is thought to be mechanically stabilizing, thus protective. Recent studies showed that sphingolipids mediate signaling cascades, involved in proliferation, apoptosis and stress response (Spiegel and Milstien, 2002; Hannun and Obeid, 2002). They are also thought to be involved in the formation of the lipid microdomains, or 'lipid rafts', proposed in 1997 by Kai Simons (Simons and Ikonen, 1997).

The gangliosides as the most complex sphingolipids contain as hydrophilic head group an oligosaccharide chain with one or more acidic sugar molecules. The acidic sugar molecule is sialic acid or N-acetylneuraminic acid (NANA or NeuNAc) whose position and number can be used to sort the gangliosides into different classes. One distinguishes between mono- di and tri-sialotetrahexosyl gangliosides. In the case of GM1 (mono-sialo ganglioside), five monosaccharides are linked to ceramide: one glucose molecule, two galactose molecules, one N-acetylgalactosamine molecule and one NANA molecule form the complex head group. Gangliosides, for they are only found in the outer leaflet with their fatty acid chains embedded in the plasma membrane and the oligosaccharides facing the extracellular environment, play an important role in cell signaling, cell recognition and communication but also mediate bacterial toxin uptake by the cell, for instance cholera toxin specifically binds to GM1 (Fishman, 1982).

Cholesterol

Cholesterol belongs to the class of sterols and is required to establish membrane permeability, stiffness, fluidity in the plasma membrane. It is also a precursor molecule for steroid hormones and fat soluble vitamins. Its hydroxyl group interacts with the polar head groups of the membrane phosphoglycerolipids and sphingolipids, while its bulky steroid and the hydrocarbon chain are embedded in the lipid phase of the membrane, alongside the nonpolar alkyl chains of the other lipids. Cholesterol reduces the permeability of the plasma membrane to protons and sodium ions (Haines, 2001). Within the cell membrane, cholesterol also functions in intracellular transport, cell signaling and nerve conduction. Moreover it is thought to play an important role in the arrangement of membranes into macro- and microdomains (Mukherjee and Maxfield, 2004; Simons and Vaz, 2004). Cholesterol is essential for life, and is primarily synthesized *de novo* within the body with smaller contributions from the diet.

1.3.2 The 'Raft' Controversy

Since the term 'raft' was introduced in 1997 by Kai Simons (Simons and Ikonen, 1997) an avalanche was set off and many biochemical and biophysical approaches attempted to prove or confute their existence in both, native and artificial membrane models. Lipid 'rafts' can be seen as transient phase separations in the fluid lipid bilayer, where sphingolipids and cholesterol get concentrated. Due to an conglomeration of lipids containing unsaturated fatty acids, these domains are denser packed and thicker. They are thought to mediate protein accommodation and organization to mediate transport via vesicles or interaction with other proteins diffusing in the membrane. Figure 1.5 which was slightly modified after Ageirsson shows the possible organization of lipids in 'rafts'. Sphingolipids and gangliosides form together with cholesterol a liquid ordered phase (L_o) which is surrounded by other phosphoglycerolipids forming the gel like phase, the liquid crystalline phase (L_c).

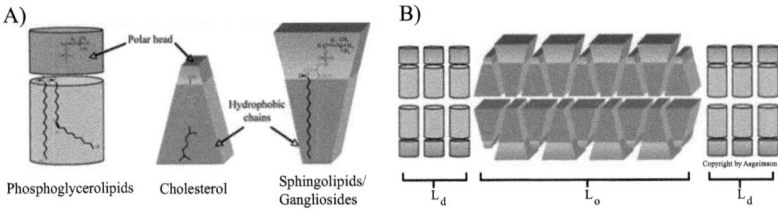

Figure 1.5: Organization of lipid 'rafts'. A shows the lipids schematically drawn: phosphoglycerolipids as cylinders, cholesterol as triangle and the sphingolipids and gangliosides as triangle standing on its tip. **B** The cylindrical phosphoglycerolipids form the liquid crystalline phase, whereas the 'raft' itself is formed by cholesterol densely packed with sphingolipids and gangliosides.

One wide-spread method of studying these microdomains relies on the distinct property that parts of the plasma membrane are detergent resistant (e.g. Triton X-100). These DRMs (Detergent Resistant Membranes) were also found to be enriched in cholesterol and sphingolipids but not in glycerolipids, leading to the hypothesis that those detergent resistant parts form small domains in the plasma membrane (Simons and Ikonen, 1997; Dietrich et al., 2001; Edidin, 2003). However, in 2002 Heerklotz *et al.* showed that this method was prone to artifacts for Triton itself affects the formation of these domains (Heerklotz, 2002). Therefore it is questionable of how meaningful DRMs are in physiological conditions.

Studies using artificial membranes as model systems showed that bilayers separate into different phases upon application of cholesterol (20%-30%) and sphingolipids (Thewalt and Bloom, 1992; Bacia et al., 2004). These domains were found to have sizes ranging from the nanoscale (de Almeida et al., 2005; Hsueh et al., 2005) to the microscale (Bacia et al., 2004) depending on the lipid composition. However, domains of these size were only found in artificial membranes and not in cell membranes although the cholsterol in the plasma membrane is comparable (30%-40%). Another drawback in the use of artificial model membranes is the lack of membrane proteins which regulate and modify membranes in living cells (Yethiraj and Weisshaar, 2007).

For investigations of 'rafts' *in vivo* common fluorescence based techniques were used, such as FCS (Bacia et al., 2004; Wawrezinieck et al., 2005), FRET (Zacharias et al., 2002) and FRAP (Feder et al., 1996; Yechiel and Edidin, 1987), or methods based on single molecule tracking (Saxton and Jacobson, 1997; Schütz et al., 2000). These studies are not described in detail here, but summarized in short. The results are contradictory, ranging from heavily clustered protein-lipid domains up to totally randomly distributed domain components. The diameters found for microdomains are also widely distributed between 25 nm and 700 nm, which can be merely due to the different spatial resolution used in each of these studies (Jacobson et al., 2007).

1.4 Proteins Interacting with Constituents of the Plasma Membrane

The interaction between the peripheral plasma membrane and the underlying cytoskeleton are crucial for migration and maintaining the shape of animal cells. It only started few years ago, when the contribution of cholesterol in the interaction of the plasma membrane with the cytoskeleton was revealed. Most membrane proteins are found in cholesterol-enriched regions that require actin anchors for maintaining their proper composition and distribution (Lillemeier et al., 2006). Very recently, it was shown that membrane cholesterol directly regulates actin stress fiber formation (Qi et al., 2009). Indeed, actin filaments are highly concentrated at the cell periphery and are part of the cell cortex. The cell cortex functions as a mechanical support of the plasma membrane. Another component of the

Introduction

cell cortex is spectrin, an actin binding cytoskeletal protein that lines the intracellular side of the plasma membrane of many cell types. It forms a scaffolding and plays an important role in maintenance of plasma membrane integrity and cytoskeletal structure. Both, spectrin proteins and actin microfilaments are attached to transmembrane proteins by linker proteins.

1.5 The Human Neuroblastoma Cell Line SH-SY5Y

The human neuroblastoma cell line SH-SY5Y is the third generation derived from SK-N-SH. The original cells were isolated from a bone marrow biopsy of a four-year old girl with metastatic neuroblastoma in the 1970s (Biedler et al., 1973). The cells' morphology is epithelial-/neuronal-like and they grow in both, monolayers and clusters. The cells propagate via mitosis and extension of neurites into the surrounding area. The SK-N-SH parental line comprises two morphologically and biochemically distinct phenotypes: neuroblastic (N-type) and substrate adherent (S-type), which can undergo transdifferentiation (Ross et al., 1983). Although derived from a neuroblastic subclone, the SH-SY5Y line retains a low proportion of S-type cells, which can be partly due to a third phenotype, the morphologic intermediate type (I-type), for this type gives rise to both N- and S-type cells 3.1. Differentiated cultures contain human dopaminergic neuronal cells presenting many characteristics of primary cultures of neurons (Encinas et al., 2000), thus providing a useful tool to perform light microscopy and biochemical large-scale studies on human neuronal cells due to an unlimited amount of cells.

1.5.1 Redifferentiation of SH-SY5Y

In agreement with previous reports (Pahlman et al., 1984), in cultures treated with retinoic acid (RA) a considerable proportion of N-type cells undergo differentiation to a more neuronal phenotype by neuritic outgrowth. On the other hand, the larger S-type cells show no apparent morphological changes, but overgrow after more then ten days of culture in the presence of RA the cultures in the subsequent days (Encinas et al., 2000).

This side effects make long-term RA treatment of SH-SY5Y unsuitable for the acquisition of homogeneous populations of differentiated cells with neuronal characteristics. However, David R. Kaplan *et al.* reported that RA induces the expression of TrkB in SH-

SY5Y cells, making them responsive to BDNF (Kaplan et al., 1993). Moreover, it has also been shown that BDNF enhances the differentiating effects of RA (Arcangeli et al., 1999). Besides the induction of TrkB expression, also LNGFR (Low affinity Nerve Growth Factor Receptor), also known as p75, was reported to be upregulated upon application of RA ((Ehrhard et al., 1993)).

Amongst other neuronal markers, the neurofilament triplet proteins are present in non-differentiated N-type and I-type cells (Encinas et al., 2000). Therefore, this cell line can be used for studies concerning the organization and localization of these filament proteins throughout differentiation.

1.6 The Cytoskeleton

Proteins of the cytoskeleton play an essential role in establishing and maintaining of cell shape in all tissues. The cytoskeleton is a system of filaments present in the cytosol of all eucaryotes, and as recently described also in procaryotes (Michie and Löwe, 2006). It provides flexibility, stability, motility and also drives and guides the intracellular traffic of organelles and vesicles. All these different tasks are executed by three families of filaments, whereas each type has distinct mechanical properties and dynamics: microtubules, actin filaments (microfilaments) and intermediate filaments (see figure 1.6 **A**).

The microtubules are rigid, hollow cylinders with an outer diameter of 25 nm built by 13 protofilaments consisting of the α- and β-tubulin heterodimer. These heterodimers are oriented in the same direction, resulting in a polarity ('+ end' and '- end') allowing for directed transport. Microtubules are commonly organized by the centrosome (MTOC, microtubule-organizing center).

The cellular shape is determined by actin filaments. These flexible two-stranded helical polymers of the protein actin have a diameter of 5-9 nm. They are organized into a variety of linear bundles highly concentrated in the cortex beneath the plasma membrane. It also features a polarity, important for e.g. whole-cell locomotion.

The ropelike intermediate filaments (IFs) consist of a large and heterogeneous family and provide for mechanical strength and resistance to shear stress. Their diameter is 10 nm, thus laying in between the microtubules and the actin filaments. The intermediate filaments are classified by similarity in structure and amino acid sequence: type I and II (acidic and basic keratins), type III (e.g. vimentin), type IV (e.g. neurofilaments and

Introduction

α-internexin), type V (lamins), type VI (nestin) and unclassified IFs.

Large-scale cytoskeletal structures can last the whole lifetime of a cell or rapidly reorganize in few minutes. These high degree of dynamics is provided by a large variety of associated proteins, which also allow for correct assembly, stabilization and regulation of the cytoskeletal elements. Beyond that another set of associated proteins, motor proteins, move organelles and vesicles along the microtubules either from the '+ end' to the '- end' (dynein) or vice versa (kinesin). The motor proteins using actin filaments as tracks belong to the family of myosins.

1.6.1 The Neuronal Cytoskeleton

Since adult neuronal cells are post-mitotic and can last throughout the whole lifetime of the organism, certain adaptations need to be taken to guarantee a lifelong stability. As was found the organization of cytoskeletal proteins in neurons differs from non-neuronal cells in many ways. Microtubules are also composed of the same basic constituents as those in non-neuronal cells, but they are more diverse. They contain many isoforms (e.g. βIII tubulin) and several different post-translational modifications as well as a large set of microtubule associated proteins (MAPs).

In contrast to all other cytoskeletal proteins, IFs display an unusual high degree of tissue specificity and state of differentiation. However, they all share homology in a core rod domain containing multiple α-helical domains. Unlike the other IFs the neurofilament (NF) triplet proteins and α-internexin are unique for their sidearms projecting from the surface. The axon caliber is mostly determined by the NF's sidearms, that are charged upon modification and therefore repel each other (Xu et al., 1996). An IF protein rather present in developing than in mature neurons is vimentin, a type III IF protein (DiazNido et al., 1996). It is not unique to neuronal precursors but also found in a large variety of other cell types. During development from neuroblasts to fully differentiated neurons, the cytoskeleton undergoes severe changes. To establish and maintain the polarized morphology of neurons, a variety of cytoskeletal elements needs to be assembled. For instance, the protein composition in the axon differs significantly from that in the cell body and in the dendrites. Initially, a neuron possesses several neurites which resemble each other in protein composition (e.g. tau and MAP2) and length. In neurons, elaborating an axon, one neurite outgrows the others losing MAP2 and enriching tau, while some of the others are stabilized and extend becoming dendritic processes in which MAP2 accumulates (Heidemann, 1996). This specification for a particular neurite is *in vivo* determined by the environment.

As size and shape change upon neuronal differentiation, the composition of the IF proteins varies coordinately (Fliegner and Liem, 1991; Lee and Cleveland, 1996). Nestin, a type VI IF protein, is expressed in multipotent neuroectodermal cells, precursors of both neurons and glia, but soon suppressed during subsequent development. Therefore it can be used as an early marker for differentiation of precursor cells (Tohyama et al., 1992). Another early IF protein is the type III protein vimentin, which is also expressed at high levels in neuroectodermal cells and concomitant with postmitotic differentiation downregulated (Tapscott et al., 1981; Cochard and Paulin, 1984; Lee and Cleveland, 1996). α-internexin (central nervous system, CNS) and peripherin (peripheral nervous system, PNS) show a similar behavior, since both are expressed very early in neuronal development but are successively downregulated (Nixon and Shea, 1992), although some neurons maintain α-internexin throughout terminal differentiation (Pachter and Liem, 1985). The light and medium subunits of the NF proteins, NFL and NFM, appear during initial outgrowth. NFH is also expressed but to a much lower extent which increases as maturation proceeds. Phosphorylation of both subunits, NFM and NFH, occurs even later, in its full extent only in large myelinated axons. Although other type IV IF proteins, such as synemin, α-internexin and peripherin accompany the NF triplet protein and co-assemble with them. In fact, α-internexin is considered being the fourth NF subunit as was demonstrated by Yuan *et al.* (Yuan et al., 2006). However, the following subsections focus on the NF triplet proteins in particular.

1.6.2 Organization of Neurofilaments

The NF triplet proteins consist of three subunits (NFL, NFM and NFH) primarily differing in the length of their C-terminal tail domain and consequently also in their molecular weight (68-70 kDa, 145-160 kDa and 200-220 kDa, respectively, in SDS PAGE). All NF proteins share a common tripartite structure, featuring a $\sim$ 46 nm-long central α-helical rod domain, a globular N-terminal head domain and a prolonged C-terminal tail domain. For NFL the tail domain is short, whereas the tail domains of NFM and NFH are extended and contain several KSP (Lys-Ser-Pro) repeats. The serine residues of these repeats are

Introduction

highly phosphorylated *in vivo* (Julien and Mushynski, 1982, 1983).

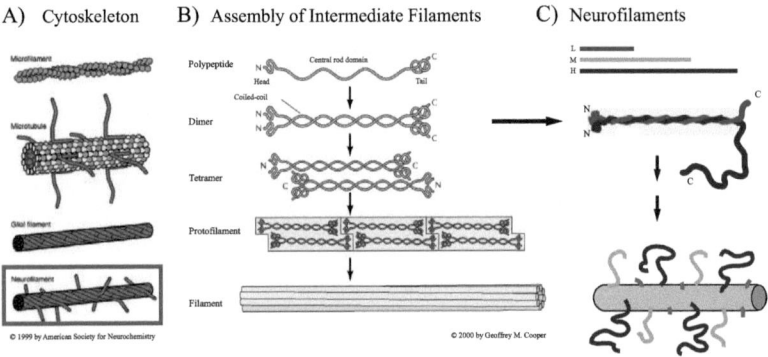

Figure 1.6: The cytoskeleton. A shows the three major constituents of the cytoskeleton: microfilaments (actin), microtubules and two members of the intermediate filaments (IFs), a glial filament and a neurofilament (NF). The assembly of IFs is shown in **B**. IFs feature a N-terminal head, a central rod and a C-terminal tail domain. They form homo- or heterodimers in a parallel side-to-side orientation. Unlike the microfilaments and the microtubules, IFs are not polar, do to the head-to-head and tail-to-tail assembly of the dimers leading to tetramers and, finally, to the 10 nm-filament. NFs consist of three subunits, the light, middle and heavy subunit (L, M and H, respectively) as drawn in **C**, which differ mostly in their length of their tail domains. They form heterodimers consisting of NFL with either NFM or NFH. In this example a heterodimer consisting of NFL (red) and NFH (blue) is shown. The long extended C-terminal tail domains are protruding from the 10 nm-filament and can be highly modified.

NF subunits proteins form heterodimers consisting of always one NFL subunit and either one NFM or NFH subunit (compare with figure 1.6 **C**). The latter ones are not able to assemble in the absence of NFL. On the other hand, NFL is capable of forming homodimers leading to short filaments without the characteristic protrusions (Geisler and Weber, 1981). As shown in figure 1.6, the first step in NF generation is to form parallel side-to-side coiled coil heterodimers. Two dimers line up in antiparallel half-staggered tetramers. These tetramers are then combined to form protofilaments, which then finally assemble to constitute the final 10 nm-filament. The C-terminal tail domains of NFM and NFH protrude from this 10 nm-filament, defining the interfilament spacing, whereas the N-terminal head and tail domains of NFL and NFM promote the lateral interactions of the protofilaments. IFs lack polarity in contrast to microtubules and actin filaments. Also opposing to the other cytoskeletal proteins they do not contain a binding site for a nucleoside triphosphate.

1.6.3 Post-translational Modifications of Neurofilaments

The best-documented post-translational modification of NFs is the phosphorylation, regarding subunit assembly and transport. Moreover, aberrant phosphorylation is a pathological hallmark of many human neurodegenerative disorders (see subsection 1.6.4). The second most abundant post-translational modification of NF is glycosylation in the form of O-linked-β-N-acetylglucosamine (O-GlcNAc) and, though to a much lower extent, also glycation, that both occur on serine and threonine residues.

Phosphorylation of Neurofilaments

NF proteins are the most extensively phosphorylated proteins in neurons (Pant et al., 2000; Grant and Pant, 2000). Most of these sites are in the tail domains of NFM and NFH but some are also found in the head domains. Phosphorylation of head domains of NFL and NFM by PKA, PKC and PKN (protein kinase A, C and N, respectively) arises mainly in the cell body shortly after synthesis, preventing the assembly of NFs or leading to their disassembly (Sihag et al., 1999; Mukai et al., 1996; Hisanaga et al., 1990a) while the tail domains of NFM and NFH (mostly on their KSP repeats) are found to be phosphorylated later, often coinciding with their entry into the axon (Nixon et al., 1989). Many contradictory roles have been attributed to the phosphorylation of the KSP repeats, including the formation of cross-bridges between NFs or with microtubules, the establishment and expansion of the axonal caliber, the slowing of the NF axonal transport and the integration of NF into a stationary network (Hisanaga et al., 1990b; Glicksman et al., 1987; Nixon et al., 1982; Lewis and Nixon, 1988; Archer et al., 1994; Jung et al., 2000; Ackerley et al., 2003; Nixon and Logvinenko, 1986).

Phosphorylation of KSP sites is carried out by two families of Pro-directed kinases: the cyclin dependent kinase CDK5 and mitogen activated protein (MAP) kinases. Several, sometimes contradictory, publications report on different possibilities to trigger the signal transduction cascades leading to an activation of these kinases: activation by growth factors (Li et al., 1999a; Pearson et al., 2001), by Ca^{2+} influx (Li et al., 1999b), integrins (Li et al., 2000) and myelination (Nixon et al., 1994). CDK5 preferentially phosphorylates KSPXK motifs of NFH both, *in vitro* and *in vivo* (Guidato et al., 1996; Sun et al., 1996), whereas the majority of KSP repeats in NF tail domains are phosphorylated by MAP kinases, as shown for mice and rat NFs. ERK1 and ERK2 (extracellular signal-regulated kinases 1 and 2) phosphorylate KSPXXK and KSPXXXK motifs on NFH (Veeranna et al., 1998), SAPK (stress activated protein kinase) acts on KSPXE motifs of NFH in the cell body upon stress (Brownlees et al., 2000), p38 MAPK (Ackerley et al., 2004; Sasaki et al., 2006) and JNK1/3 (c-Jun N-terminal kinase 1 and 3) (Brownlees et al., 2000; Giasson and Mushynski, 1996; O'Ferrall et al., 2000) phosphorylate several KSP sites, and GSK3 (glycogen synthetase kinase 3) was found to phosphorylate some KSP repeats on NFM (Guan et al., 1991) and few sites on NFH (Guidato et al., 1996). Sharma *et al.* showed that CDK5 plays an important role in human NFH tail domain phosphorylation (Sharma et al., 1999).

Moreover, Zheng *et al.* showed that phosphorylation of the head and tail domains are intimately related (Zheng et al., 2003) by an orchestrated action of PKA and MAP kinases. NF head domain phosphorylation inhibits NF assembly and C-terminal phosphorylation in the cell body, thus protecting the neuron from abnormal accumulation of phosphorylated NF in the perikaryon and allowing for transport into and along the axon. NFs lose to a great extent their ability in forming interconnected networks upon dephosphorylation. This removal of the phosphate group is mainly (60%) catalyzed by PP2A (protein phosphatase 2A) (Veeranna et al., 1995; Saito et al., 1995) and to a lower extent (10-20%) by PP1 (protein phosphatase 1) (Strack et al., 1997).

O-GlcNAcylation of Neurofilaments

O-GlcNAcylation, unlike phosphorylation, is carried out by only one enzyme, O-linked-β-N-acetylglucosamine transferase (OGT) which catalyzes the transfer of β-N-acetylglucosamine (GlcNAc) from UDP-GlcNAc to serine and threonine residues of proteins. OGT is very sensitive to relatively small changes in substrate availability over a wide range of concentrations (Kreppel and Hart, 1999), meaning that its activity is regulated by the intracellular glucose metabolism via alteration of UDP-GlcNAc production through the hexosamine biosynthesis pathway (HBP). Most of the glucose in the brain is metabolized to produce ATP. However, 2-5% attain the HBP to produce glucosamine-6-phosphate, and, finally, UDP-GlcNAc (Love and Hanover, 2005). The enzyme removing the O-GlcNAcylation is O-GlcNAcase (OGase).

O-GlcNAcylation is a common modification of cytosolic and nuclear proteins that regulates protein stability, subcellular localization and interaction between proteins (Slawson and Hart, 2003). The function still remains elusive, but several studies suggest that these modifications play a role in NF assembly. O-glycosylation could reciprocally modulate its phosphorylation and therefore also influence the assembly and dynamics of NF (Perrot et al., 2008). With the production of highly specific antibodies against the these modifications, e.g. a monoclonal antibody recognizing an O-glycosylated epitope of the NFM tail domain (Ludemann et al., 2005)

Both phosphorylation and O-GlcNAcylation occur on similar serine and threonine residues (Dong et al., 1993). Therefore it may be possible that both modifications compete for these sites. They may furthermore also alter the substrate specificity of nearby sites by steric or electrostatic effects.

1.6.4 Role in Neurodegeneration

A hallmark of several human neuropathologies is the abnormal agglomeration of NFs, such as the accumulations in the cell bodies of motor neurons in amyotrophic lateral sclerosis (ALS) (Kato and Hirano, 1991), in Lewy bodies of Parkinson's Disease (Hill et al., 1993), and the neurofibrillary tangles of Alzheimer's Disease (Ulrich et al., 1987). These accumulations can be potentially induced by multiple factors, such as dysregulation of NF expression, mutations, defective axonal transport and abnormal PTMs.

The neurofibrillary tangles (NFTs) of Alzheimer's Disease consist of tau, NF and other cytoskeleton proteins (Ishii et al., 1979). Still, NFT formation is not completely understood as well as their role in the curse of the disease. But one of their characteristics is besides the tau hyperphosphorylation an extensive NF phosphorylation (Sternberger et al., 1985; Wang et al., 2001). Furthermore, it is known that during Alzheimer's Disease glucose brain metabolism is impaired (Perry et al., 2003; Iqbal and Grundke-Iqbal, 2005). Moreover, Deng *et al.* (Deng et al., 2008) suggested that the hyperphosphorylation and agglomeration might be a cause of this impairment. Also in other neurodegenerative diseases, for instance ALS or Parkinson's Disease, the phosphorylation state of NF subunits is misregulated (Forno et al., 1986; Manetto et al., 1988).

Specific labeling allows the investigation of the different post-translational modified epitopes of neurofilament proteins. However, only severe changes in the PTM pattern can

Introduction

be revealed by conventional optical microscopy. Therefore, there is the need for higher resolution to detect early alterations.

2 Motivation

The aims of this thesis were to apply the STED (STimulated Emission Depletion) concept in imaging and spectroscopy to investigate alterations occurring on the protein and lipid level.

In the first part the organization of the two major post-translational modifications (PTMs), phosphorylation and glycosylation on neurofilaments is examined. Neurofilaments are one of the main constituents of the neuronal cytoskeleton and are highly modified by phosphorylation and glycosylation. An imbalance between these two modifications is a pathological hallmark in many age-related neurodegenerative diseases, such as Alzheimer's Disease. Since the neurofilaments and their modifications are in a spatial range < 40 nm, which lies far beyond the diffraction limit of conventional light microscopes (~ 250 nm), the organization of different PTMs could so far not be detected by light microscopy. The ~ 10-fold higher resolution of the STED microscope allows the direct investigation of both modifications on neurofilaments. One severe effect of this imbalance between both modifications, can result in the formation of axonal swellings and finally neurodegeneration. There is the hope, that, with the higher resolution of the STED microscope alterations in the pattern can be detected to an earlier point in time.

In the second part, the diffusion behavior of phospholipids is investigated in the plasma membrane of living cells. The properties and constitution of the eucaryotic plasma membrane is changing upon senescence of the cell. Therefore, it is important to gain more insight into the diffusion characteristics of the lipids to understand downstream signaling effects. These characteristics are conventionally investigated by FCS (FLuorescence Correlation Spectroscopy). Since fluorophore conjugated lipids are required for FCS experiments, unwanted side effects by the dye have to be ruled out to ensure proper membrane anchoring of the lipid. A term that led to a lot of discrepancy in the scientific community during the last years, is the term 'raft'. Many different techniques had been used to describe the nature of these microdomains. The results are contradictory, ranging from

Motivation

heavily clustered protein-lipid domains up to totally randomly distributed domain components. Also the description of their spatial extent greatly varies between 25 nm and 700 nm. Applying the STED concept also to FCS, enables a seamless decrease of the focal volume from $\sim 250-40$ nm, thus resulting in a more precise detection in the spatial dimension. Furthermore, a change occurring during aging in the plasma membrane is the alteration of the endogenous cholesterol level. Sphingolipids unlike phospholipids are thought to interact with cholesterol enriched domains. The usage of STED-FCS for the investigation of the diffusion behavior of phospholipids could lead to more insight.

3 Results

Aging or senescence of organisms and single cells can be characterized by their declining ability to respond to stress, increasing homeostatic imbalance and increased risk of aging-associated diseases, such as neurodegenerative diseases, with cell death usually being the ultimate consequence. Under normal conditions, aging appears to be irreversible and requires morphological and biochemical changes of the cell.

With proceeding aging, neuronal cells become susceptible to neurodegenerative diseases. A frequently observed accompanying effect is the pathologic accumulation of hyperphosphorylated proteins, for instance, neurofilament subunits. These agglomerations impede axonal transport and finally lead to neurodegeneration. Perturbations in the metabolism also occur during aging, such as diabetes type II.

Another aspect in aging is the alteration of plasma membrane composition (Venable et al., 2006) and therefore its fluidity. Especially cholesterol and ceramide levels are altered, which both play major roles in the establishment of lipid microdomains. Since these domains are known to mediate various signaling cascades (Tsui-Pierchala et al., 2002), investigations of the properties of the lipid microdomains could deliver deeper insight to understand the downstream effects in aging.

The following sections report on the investigation of post-translationally occurring modifications (such as phosphorylation) of neurofilaments in the background of metabolic stress and on the characterization of lipid nanodomains applying the STED concept.

3.1 Post-translational modifications of neurofilaments visualized by STED Microscopy

In the first part of the thesis, the post-translational modifications (PTMs) of neurofilaments were examined by imaging using STED (stimulated emission depletion) microscopy, a high resolution light microscopy method, overcoming the diffraction limit.

To investigate the PTMs of the NF subunits a suitable model system is required. The human neuroblastoma cell line SH-SY5Y is a cell culture model that circumvents the need of primary cultures or brain slices, as such providing a high reproducibility and easier handling.

The proteins, which took center stage in the first part of the thesis, the neurofilament triplet proteins, are conserved throughout mammals. However, the primary structure of NF heavy subunit, for instance, indicates that the kind and number of kinase binding sites that become phosphorylated greatly vary between human, mouse and rat. A widely investigated kinase involved in normal phosphorylation of NFs is CDK5 (cyclin dependent kinase 5). In the tail domain of the heavy neurofilament subunit of humans, there are 43-44 CDK5 binding sites, whereas in mouse and rat only 9-10 are present (Elhanany et al., 1994; Jaffe et al., 1998). This dissimilarity also indicates a discriminative regulation of phosphorylation and possibly also other PTMs in different organisms by a divers set of enzymes. The SH-SY5Y model system allows the investigation of the PTM organization in a human background which can be used to investigate particular conditions found in neurodegenerative diseases affecting humans.

SH-SY5Y consists of three distinct cell types: the substrate adherent S-type, the neuroblastic N-type, and the morphological intermediate I-type, which, as a multi potent stem cell, is the progenitor of the two other cell types. In their native state, these cells express the NF subunit proteins (L, M and H) only to a low extent. However, SH-SY5Y cells can be redifferentiated which leads to the expression (or increase of expression, respectively) of all three neurofilament protein subunits. Therefore, at first, fully redifferentiated cultures had to be generated. The appearance and disappearance of certain IFs could be used as an indicator for reaching different stages of differentiation.

3.1.1 Appearance of Intermediate Filament Proteins during Redifferentiation of SH-SY5Y

There are different possibilities which are known to induce redifferentiation in neuroblastoma cells. Figure 3.1 highlights the different stages of divers redifferentiation methods considering cell morphology and intermediate filament (IF) expression.

As already shown previously by Encinas *et al.*, the heterogeneous cell line SH-SY5Y

3.1.1 Appearance of Intermediate Filament Proteins during Redifferentiation of SH-SY5Y

(see figure 3.1 **Step 1** and **Step 2**) could be sensitized for brain derived neurotrophic factor (BDNF) (Encinas et al., 2000). The treatment with all-*trans* retinoic acid (RA) promotes the N-type cells (see figure 3.1 **Step 3**) and also leads to their sensitization for BDNF by upregulation of TrkB, a member of the tyrosine kinase receptors, and to a lower extent also LNGFR (low affinity nerve growth factor receptor, also called p75). A subsequent application of BDNF finally gives rise to fully differentiated human neuron-like cells (Ehrhard et al., 1993; Encinas et al., 2000) as shown in figure 3.1 **Step 4a**

Other methods to induce redifferentiation in neuroblastomas are the application of 12-tetradecanoyl-13-acetyl-β-phorbol (TPA), a protein kinase C (PKC) activator (Troller et al., 2001), and on the other hand the deprivation of serum from the culture medium (Glick et al., 2000). Both methods also resulted in neurite outgrowth and the expression of neuronal markers in N-type cells after several days. The network formation between the cells was not as pronounced as after the sequential treatment with RA and BDNF as can be seen in figure 3.1 **Step 4b**. Moreover, the survival of especially the serum deprived cells was reduced from a few days up to one week in maximum. The TPA differentiated N-type cells were either overgrown by the S-type cells when cultured in the presence of serum or could not survive for more than few days, when grown without serum.

Therefore, all following redifferentiation experiments were performed treating the cells sequentially with RA and BDNF, respectively.

An IF protein present in non differentiated SH-SY5Y cells is vimentin (highest in S-type, lowest in N-type cells) as shown in figure 3.1, whereas the NF triplet proteins and α-internexin were exclusively expressed in N- and I-type cells, respectively. Since the I-type, as the progenitor of the other two cell types, expresses markers of both, N- and S-type cells, it was not considered in the following. NF triplet proteins are predominantly localized in the axon and in the perikaryon and are not found in dendrites. In non differentiated cells, lacking axons, they are predominantly localized in the cell body of N-type cells, playing a minor role in establishing and maintaining cell shape. Upon gradual redifferentiation of SH-SY5Y, the N-type cells were promoted induced by the application of all-*trans* retinoic acid (RA) for 5-7 days. As shown in figure 3.1 **Step 3**, neurite outgrowth can be observed by staining of NF subunits (here, this step is referred to as differentiation stage I). Note that the S-type cells are still present in the culture at that stage (after 5 days in the presence of RA) but show no immunoreactivity towards the pan-NF antibody since they lack NF proteins. A prolonged application of RA (>9 days) led to an increased

Results

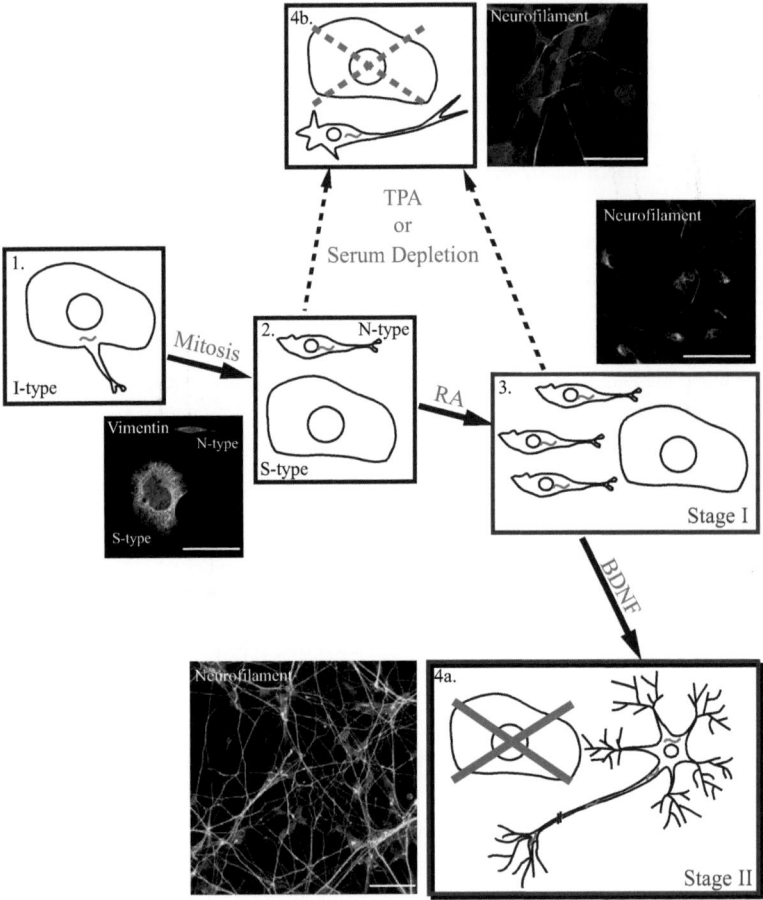

Figure 3.1: Scheme of the redifferentiation of the human neuroblastoma SH-SY5Y. The stylized red filament indicates the expression of neurofilament proteins. **Step 1**: The stem cell like I-type cell is the precursor of the non-neuronal S-type and the neuronal N-type cell (**Step 2**). Note the size difference between both cell types in the immunofluorescence image. **Step 3**: Upon the application of retinoic acid the N-type is promoted. Here, this stage is referred to as differentiation stage I. **Step 4a**: After incubation with hBDNF in the absence of serum, strong neurite outgrowth and network formation can be observed, referred to as differentiation stage II. Incubation of cells in serum deprived medium without BDNF, or the incubation with TPA, also causes neurite outgrowth but network formation to a lower extent (**Step 4b**). The length of the scale bars is 50 μm.

propagation of the S-type cells, finally overgrowing the N-type cells (not shown). The application of BDNF in the absence of serum subsequent to the RA treatment for 5-7 days finally led to the formation of fully redifferentiated N-type cells after several days, which survived for up to several weeks (see figure 3.1 **Step 4a**). These cultures showed a strong immunoreactivity for the pan-NF antibody with a now predominant axonal NF localization rather than the cell body. At that stage (here referred to as differentiation stage II), most S-type cells were starved to death due to the absence of serum. The NF proteins subordinated the role of vimentin with ongoing differentiation. The next subsection further clarifies the decreasing role of vimentin throughout redifferentiation.

3.1.2 Reciprocal Expression of Vimentin and Doublecortin

A pivotal indicator for precursor neurons is the 43-53 kDa microtubule-associated protein doublecortin (DCX) which is exclusively expressed in immature neurons. It is found in cell bodies and leading processes of migrating neurons and axons of differentiating postmitotic neurons.

Western blot analysis of the expression level of vimentin in the course of differentiation clearly shows a more or less constant level until differentiation stage I as was already seen by Sharma *et al.* (Sharma et al., 1999). This level suddenly drops at differentiation stage II after the cells had grown in the presence of BDNF for several days. An opposite behavior can be observed for the expression level of DCX. At differentiation stage II strong bands are observable at $\sim$44 kDa, coinciding with the loss of vimentin (see figure 3.2 **A**).

The protein distribution throughout the different cell types, however, cannot be revealed biochemically. Therefore immunofluorescence was carried out to investigate the cell type specific expression pattern. A co-staining of vimentin and DCX at differentiation stage I, II and in non differentiated cells highlighted a reciprocal proportion of the two proteins (see figure 3.2 **B**). Vimentin formed nice networks in S-type cell bodies whereas in non differentiated N-type cells the filaments appeared less prominent. This impression was further enhanced for N-type cells at differentiation stage I, where the vimentin network seemed to be rather fragmented, up to only a tiny residual meshwork left in the perikaryon of N-type cells at stage II. This observation led to the conclusion that mostly S-type cells contributed to the bands observed for vimentin in figure 3.2 **A**. Since

Results

only BDNF-sensitized N-type cells could survive in the absence of serum, the S-type cells were starved, thus leading to the observed decrease in the expression level of vimentin in the western analysis (indicated by the black arrow).

Coinciding, DCX was exclusively present in N-type cells, in diffuse traces already at differentiation stage I. However, shortly after N-type cells having exited the cell cycle, a strong increase of DCX could be detected at differentiation stage II (see figure 3.2 **B**). Cells expressing DCX were lacking vimentin which is shown by the either only red or green but never yellow (co-expression of DCX and vimentin in the same cell) staining of the cells. Presumably, at that stage DCX is associated with the microtubules, as was concluded from the filamentous appearance of the DCX immunostaining.

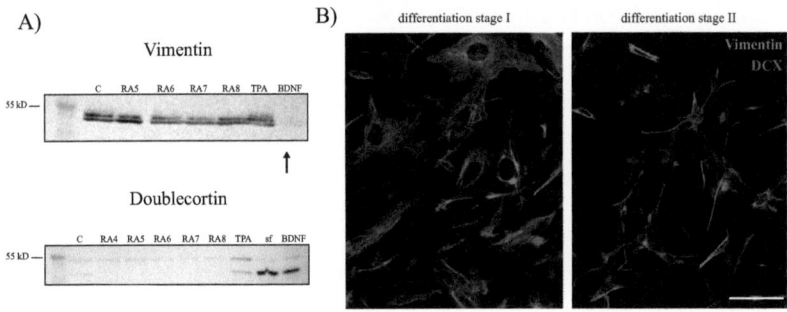

Figure 3.2: Vimentin and doublecortin throughout redifferentiation. Vimentin is expressed throughout the heterogeneous SH-SY5Y culture but plays a subordinate role when the cells progressively develop during redifferentiation, as highlighted in **A**. The expression level of vimentin suddenly drops down in cells at differentiation state II. Coincidental, DCX appears with a strong band in BDNF and serum-deprived cells and with a weak band also in TPA differentiated cells. In **B** an immunocytochemical double staining of vimentin (green) and DCX (red) is shown in cells at differentiation stage I and differentiation stage II. The scale bar is 50 μm.

3.1.3 Visualization of Post-Translational Modifications of Neurofilaments using STED Microscopy

Neurofilaments are extensively modified on both the C-terminal tail and the N-terminal head domains. The best-documented PTM (post-translational modification) of NFs is the phosphorylation carried out by second-messenger dependent and second-messenger independent kinases. The attachment of a phosphate group to either a serine or a threonine residue is highly dynamic due to the action of a different set of kinases and phosphatases. Another frequently occurring PTM is the glycosylation which like phosphorylation is also dynamically regulated.

Monoclonal antibodies (SMI, Sternberger Monoclonals Incorporation) were used to detect the modified epitopes. SMI34 detects the phosphorylated epitope of NFH and to a lower extent also of NFM. The dephosphorylated epitopes of NFH and to a much lower extent also of NFM are recognized by SMI33. SMI36 was used to detect the phosphorylated epitope of NFH, whereas the O-GlcNAcylated epitope of NFM was detected by NL6. All epitopes investigated here were located at the tail domain of the NF subunits. The high resolution STED images were recorded using a supercontinuum STED setup (described in subsection 5.6.4). Since the excitation and the STED wavelengths were generated by the same supercontinuum laser source, this setup could be operated at different wavelengths pairs of excitation and STED wavelengths. Here, the visible organic dye Atto590 performed best and was therefore conjugated to the secondary antibodies (see subsection 5.5.2). The excitation wavelength was chosen to be 570 nm and the accordant STED wavelength 700 nm. The STED and confocal reference images were recorded sequentially, taking the STED image (pixel size 15 nm; STED resolution $\sim$30 nm) always before the confocal image (pixel size 40 nm; confocal resolution $\sim$250 nm).

As depicted in figure 3.3, the three different stainings for the phosphorylated, the dephosphorylated and the O-GlcNAcylated epitopes cannot be distinguished by confocal microscopy. However, there is some evidence that the phosphorylated epitopes appear less continuously than the O-GlcNAcylated epitopes. The higher resolution of the STED microscope is necessary to reveal the clearly punctuated phosphorylated epitopes and the more regular and homogeneously distributed staining of NL6. Moreover, it can be seen that the dephosphorylated epitopes have a similar organization pattern as the O-GlcNAcylated sites and different from the phosphorylated epitopes.

3.1.4 Hyperphosphorylation upon Glucose Deprivation

The fact that the organization pattern of different PTMs could be finally revealed can be furthermore used to study pathological phenomena, such as the hyperphosphorylation of neurofilaments. This hyperphosphorylation is a frequently accompanying phenomenon

Results

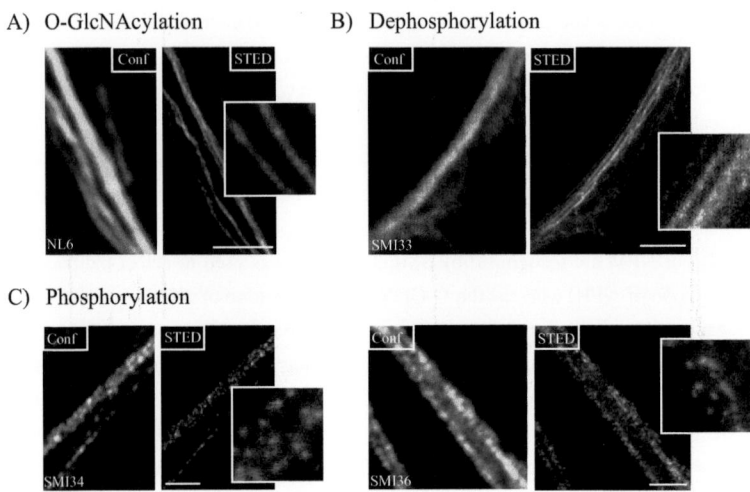

Figure 3.3: Post-translational modifications of neurofilaments. Three different examples are shown using antibodies specifically recognizing the modified epitopes. **A** As an example for glycosylation the O-GlcNAcylated epitope of NFM is detected by NL6, SMI33 highlights the dephosphorylated epitopes of both NFH and to a lower extent NFM (**B**), whereas SMI34 and SMI36 specifically detect the phosphorylated epitopes of both NFH and to a lower extent NFM, and only NFH, respectively (**C**). All images were taken in confocal and STED modes. Having a closer look at the confocal images, only few details about the organization of these modifications are visible since most features are blurred. The insets of the STED images reveal details such as the more or less continuous distribution of the glycosylated epitope, which is less pronounced for the dephosphorylated epitopes, and is clearly spotted for the staining with SMI34 and SMI36. The pixel size is 15 nm for the STED images and 40 nm for the confocal images. The size of the bar in all images is 2 μm and the box length of the insets is 1 μm.

found throughout aging and neurodegenerative diseases. As was suggested by Perrot *et al.*, phosphorylation and O-GlcNAcylation could be reciprocal modifications at the same sites of the proteins (Perrot et al., 2008). Moreover, since there is a lot of evidence that the glucose metabolism is impaired in the course of several neurodegenerative diseases (Perry et al., 2003; Iqbal and Grundke-Iqbal, 2005), the hyperphosphorylation could be a direct (or indirect) effect of this impairment.

To study the hypothesis whether the glucose metabolism has a direct effect on the balance between O-GlcNAcylation and phosphorylation in this here used human neuroblastoma cell culture model, the cells that were fully redifferentiated were deprived

3.1.4 Hyperphosphorylation upon Glucose Deprivation

of glucose. This was simply realized by replacing the differentiation medium II with a deprivation medium lacking glucose (see table 5.5 for medium composition). After incubating the cells for various durations ranging from several hours up to several days in that medium, they were immunolabeled using monoclonal antibodies directed against the phosphorylated (SMI34 and SMI36, respectively) and against the O-GlcNAcylated (NL6) epitopes.

Since the pattern of different PTMs can differ along the length of the axon, more than three areas of the same axon were imaged and compared. Indeed, as can be easily seen by wide field microscopy, O-GlcNAcylation is found more abundant close to the cell body than towards the end of the axon, while phosphorylation behaves opposite. However, the type of pattern, punctuated versus continuous, remained the same.

Severe changes, such as axonal swellings, could be observed by standard confocal microscopy after several days in the absence of glucose, accompanied by increased cell death. These axonal swellings were positive for the staining of the phosphorylated epitopes by SMI34 and SMI36 (see figure 3.4 **D**). On the other hand, the staining of the O-GlcNAcylated epitopes was strongly reduced (see figure 3.4 **H**). That indicates a misregulation between phosphorylation and O-GlcNAcylation which became apparent for conventional light microscopes only after several days in the absence of glucose. However, the question arose whether forerunners of this imbalance are already detectable at earlier time points. Since confocal microscopy was unable to distinguish between the two PTMs, the following experiments were also performed with the supercontinuum STED setup. The images shown in figure 3.4 reflect the typical intersection of all images taken for the different conditions and at different axonal positions.

Examining the overview images of the phosphorylated epitopes in figure 3.4 **A-D**, in both the STED and the confocal mode, no significant change could be seen except for the axonal swellings shown in **D** which occurred after several days (typically after 4 to 5 days but in some examples also already after 48 h). But having a closer look at the details of the pictures slight changes could be observed in the STED mode starting from the control in **A** throughout the ongoing deprivation for $\sim$9 h (**B**) and $\sim$24 h (**C**): the clearly punctuated organization of the phosphorylated subunits depicted in **A** was altered to a still spotty but more condensed pattern, as depicted in the inset of **B** to an almost continuous appearance in **C**. Finally, after formation of the axonal swellings, no filamentous structures but only large agglomerates could be observed in both confocal and STED mode. On the other

Results

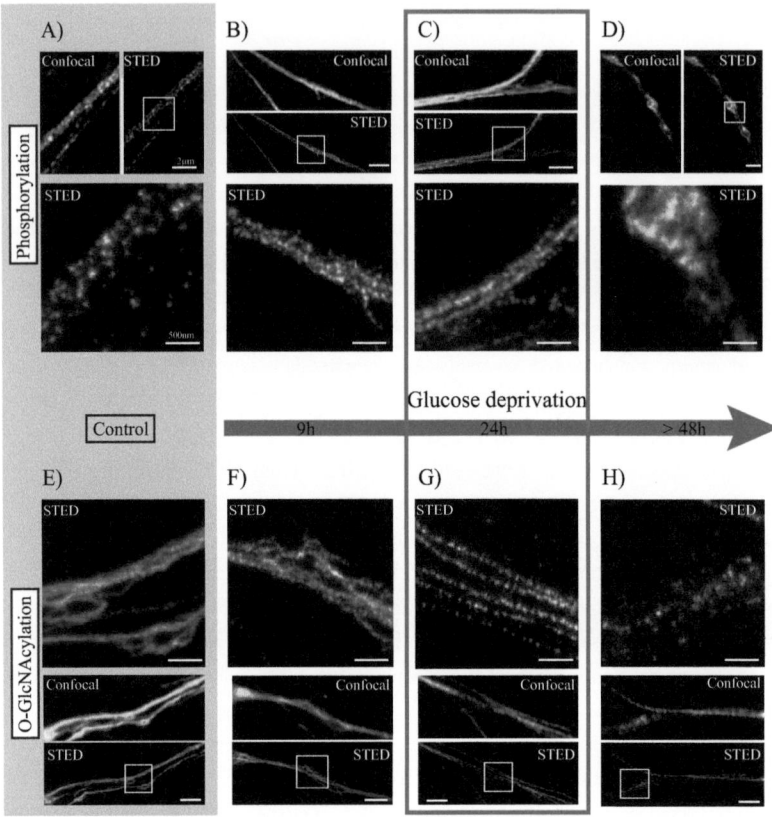

Figure 3.4: Effect of glucose deprivation on the organization of PTMs. Redifferentiated SH-SY5Y cells were glucose deprived for different times ranging from few hours up to several days and images were recorded in both STED and confocal modes. The phosphorylated (**A-D**) and the O-GlcNAcylated (**E-H**) epitopes were immunolabeled with Atto590 using indirect immunofluorescence. In **A** and **E** control images of cells grown in the presence of glucose are shown. **B** and **F** show cells which had been deprived between three and nine hours, in **C** and **G** cells are displayed which had been grown in the absence of glucose for up to 24 hours, whereas the cells shown in **D** and **G** had been deprived for several days (≥ 2). All different conditions were recorded with $n \geq 25$ cells. The size of the bar in the overview images is 2 μm, whereas the bar size in the details is 1 μm.

side, as depicted in figure 3.4 **B**, the glycosylation of neurofilaments appeared as continuous pattern (within the achieved resolution of this STED setup which was about the size of the detection complex of target, primary and secondary antibodies). Upon glucose deprivation for $\sim$9 h (**F**) to $\sim$24 h (**G**) a loss of continuity in the staining was observed, finally leading to a complete loss of filamentous appearance after glucose deprivation for several days (**H**). O-GlcNAcylated NFM was not detected in axonal swellings. The mitochondrial potential was slightly affected but did not collapse upon glucose deprivation as was checked by the potential-sensitive dye JC-1 (see subsection 5.5.1 and 4.3).

The resolution and overall performance of the microscope was constantly checked by aligning the excitation and the two STED beams by means of small gold beads in the focal plane between each exchange of the samples.

Refeeding glucose after deprivation had different effects depending on the duration of the previous deprivation. Long time deprived cells (>48 h) which had been refed with glucose for 24 h quickly died shortly after (or while refeeding), similar to the deprived cells that had not been refed. On the other hand, upon application of glucose not later than $\sim$24 h, cell death could be avoided.

First results indicated the tendency that the application of pyruvate (> two-fold molar excess of glucose) during glucose deprivation did not alter the PTM patterns observed for the glucose deprived cells but led to a prolonged cell survival. This further indicates that the application of this downstream substrate of glycolysis had little or no effect on the regulation of the PTMs of NF subunits, which is further discussed in subsection 4.1.4.

In conclusion, with only confocal resolution, alterations in the pattern of the PTMs could only be detected after several days of glucose deprivation, when the changes were too severe to prevent cell death by refeeding glucose. The higher resolution of STED microscopy enables one to reveal changes in the organization pattern of PTMs already after $\sim$24 h (see figure 3.4, green box) of deprivation when refeeding resulted in a prolonged cell survival.

3.1.5 The Role of p38 Mitogen Activated Protein Kinase in Glucose Deprived cells

The question arose what caused this hyperphosphorylation of the NFs. Several kinases are known to interact with NF proteins. PKA, PKC and PKN are responsible for the

Results

phosphorylation of the NF head domains. Since the pathologic phosphorylation occurs on the C-terminal tail domains, these kinases were not considered. On the other hand, it was shown that the mitogen activated protein kinase p38 MAPK is involved in aberrant NFM and NFH tail domain phosphorylation in ALS (Ackerley et al., 2004). Therefore, this mitogen activated kinase was examined for its putative involvement in glucose deprivation mediated phosphorylation events. To investigate its contribution, a kinase activity assay was performed based on immunoprecipitation (IP) of the active kinase from the cell lysate and the subsequent phosphorylation of its substrate *in vitro* (the protocol is described in detail in subsection 5.4.2). The activity of p38 MAPK was then determined by detecting phosphorylated ATF-2 (pATF-2, MW ~40 kDa) by western blot analysis.

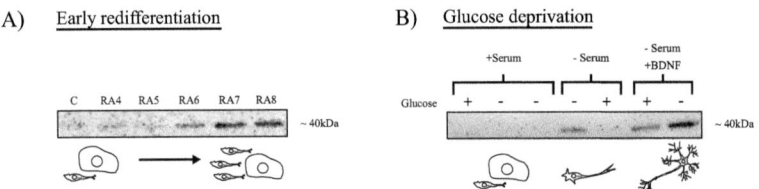

Figure 3.5: Activity assays of p38 MAPK. This assay relies on the immunoprecipitation of active (phosphorylated) p38 MAPK and the subsequent incubation with its substrate ATF-2. **A** shows the p38 MAPK activity during differentiation until stage I (application of retinoic acid, RA for 4-8 days). The morphologies of the cells are shown schematically below the blots. The effects of different starving conditions (serum and glucose) and the application of BDNF on the activity of p38 MAPK are shown in **B**. The plus or minus signs indicate the presence or absence of glucose (1 g/l) in the medium.

As a positive control, the activity of p38 MAPK was observed during early differentiation of SH-SY5Y (see figure 3.5 **A**), since this kinase is known to play an important role during differentiation (Liu et al., 2004). Cell lysates were prepared from non differentiated cells (C, control) and cells incubated in the presence of RA for various times (RA4 to RA8, incubation with retinoic acid for four to eight days). As expected, an activity increase of p38 MAPK was observed during early differentiation (stage I). In figure 3.5 **B** the effects of different starvation conditions (serum deprivation and glucose deprivation, respectively) and the completion of redifferentiation (stage II) induced by BDNF on p38 MAPK activity is shown. Non differentiated cells (top three lanes) were grown in serum containing medium and glucose deprived for different periods (9 h and 24 h).

No pATF-2 was detected independent from the deprivation duration. In cells growing in serum free medium, differentiation was induced as depicted by the sketch. However, as already mentioned earlier, in the absence of glucose their anyway limited viability was further reduced. In the presence of glucose no phosphorylated substrate was visible in the blot. However, when these cells were additionally glucose deprived for $\sim$ 24 h, pATF-2 was detectable. Fully differentiated BDNF-dependent cells were cultured in serum free medium in the presence of BDNF. In accordance with the activity enhancement upon RA induced differentiation, p38 MAPK activity is observable by a strong pATF-2 band. Its activity was further strongly enhanced upon glucose deprivation for $\sim$ 24 h. The contribution of another MAP kinase, the stress activated protein kinase c-Jun N-terminal kinase (SAPK/JNK) is reviewed in subsection 4.1.4.

3.1.6 The Role of c-Jun N-terminal Kinase in Glucose Deprived cells

Another kinase which is activated upon stress is the stress activated protein kinase (SAPK) c-Jun N-terminal Kinase (JNK). Its substrate c-Jun is a protein which, together with c-Fos, forms the AP-1 early response transcription factor. Since this kinase was found to be involved in NF subunit phosphorylation in primary rat neurons (Brownlees et al., 2000), its activity in the here used human model system was also investigated.

Two different techniques were applied, including an IP-based method similar to the p38 MAPK assay and an ELISA-based assay (the protocols are described in subsection 5.4.2 in detail). As can be seen in figure 3.6 **A**, no increase in JNK activity is observable during redifferentiation. However, upon glucose deprivation for $\sim$15 h, the phosphorylated c-Jun band at $\sim$37 kDa is stronger corresponding to an increased JNK activity. Additionally, a second, ELISA-based assay was performed by directly measuring JNK phosphorylation (activation) on cultured cells. Two 96-well plates were prepared in parallel. As a positive control, one plate was irradiated with UV light for 10 s (orange) and incubated at 37 °C and 5% CO_2 for additional 15 min, whereas the other plate was treated in the same way but without UV irradiation (green). Via colorimetric reaction and an ELISA plate reader the amount of active and total JNK was determined and normalized to the amount of cells.

As can be seen in figure 3.6 **B**, cells grown under normal conditions with glucose

Results

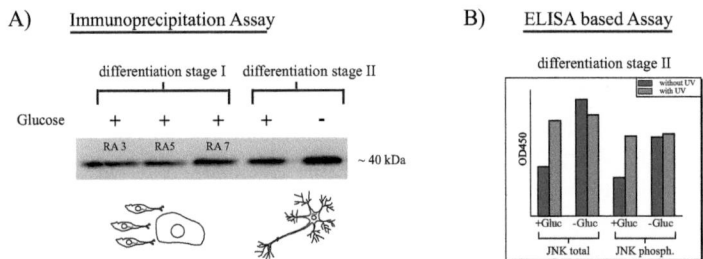

Figure 3.6: Effects of glucose deprivation on JNK activity in fully differentiated SH-SY5Y cells. In **A** an immunoprecipitation kinase assay comparable with the p38 MAPK assay is shown depicting the phosphorylated substrate phospho-c-Jun at ∼37 kDa. As can be seen, throughout differentiation (stage I and II) no alteration in activity can be monitored. However, upon glucose deprivation for ∼15 h (-), a slight increase is observable. In **B** the results of the ELISA-based assay are shown. As positive control, cells were irradiated with UV light (orange columns).

showed a twofold increase of both total and phosphorylated JNK when irradiated with UV light (orange) as compared to the not exposed cells (green). However, upon glucose deprivation, total and active JNK is increased in the non irradiated cells, reaching the same level than the positive controls. The activation of the kinase upon UV irradiation was not further enhanced by glucose deprivation.

In conclusion, both kinases p38 MAPK and JNK showed an increase in activity upon long term glucose deprivation in fully redifferentiated SH-SY5Y. Since, amongst other proteins, the tail domains of the neurofilament subunits M and H are substrates of p38 MAPK, it possibly contributes to the observed pathologic hyperphosphorylation events to a large extent.

3.2 Lipid Diffusion in the Plasma Membrane

To investigate the dynamic behavior of lipids in living cells, a technique is required that features two properties: the capability to discriminate a certain type of lipid from the heterogeneous background of the other plasma membrane lipids, and the possibility to do it in living cells. Light microscopy combined with FCS provides an ideal tool for the analysis of lipid dynamics in living cells. The major results of this section are published in (Eggeling et al., 2009).

A suitable label (e.g. a fluorophore) is required for the analysis which can specifically be detected. Detecting endogenous lipids by the use of fluorescently labeled affinity markers, e.g. antibodies, is not recommended because of the tremendous difference in molecular weight of the lipids ($M_{lipid} \approx 1\,\text{kDa}$) compared to the antibody molecules or antibody fragments ($M_{IgG} \approx 150\,\text{kDa}$ and $M_{F_{ab}} \approx 50\,\text{kDa}$), which may alter the localization and the dynamic behavior of the lipid. Furthermore one risks the formation of artificial clusters due to the tendency of cross linking by the antibodies.

To exclude possible interferences with the lipid's localization and function, a fluorophore was attached directly to a lipid and then incorporated into the plasma membrane. But, since the size of the rather small organic fluorophores ($\approx 0.6\,\text{kDa}$) is still large compared to the lipid ($\approx 1\,\text{kDa}$), the general labeling strategy was to substitute a part of the lipid (e.g. one of the acyl chains) with the dye molecule. The lipids were either labeled at their head groups, i.e., in the water phase, or at the water-lipid interface, i.e., by acyl chain replacement. To ensure proper function, the replacement of one acyl chain needed to meet some prerequisites: as the acyl chains are located in the lipid phase the dye-conjugated chain still had to be incorporated there. Further the steric properties of the lipid must not to be changed by the dye. The first requirement could be met by using the lipophilic dye Atto647N. The second criterion, however, was harder to fulfill since most of the organic dyes are bulky molecules. It was already shown previously (Wang and Silvius, 2000) that the attachment of a dye to a lipid molecule may alter its behavior. Therefore several control experiments were performed to rule out possible side effects. These side effects are reviewed in subsection 3.2.3 considering the influence of the kind of label or the labeling position on the lipid's incorporation ability, its polarity or its diffusion behavior as well as possibly induced alterations of the endogenous lipid composition.

Results

3.2.1 Reduction of the Focal Volume by Combining STED with Fluorescence Correlation Spectroscopy

In biophysics, single molecule fluorescence spectroscopy is a well established, widely used technique to study fast temporal processes and the dynamics behind them. Single molecule techniques span large time scales ranging from picoseconds to minutes (Sprague et al., 2004; Eggeling et al., 2001). The advantage of fluorescence correlation spectroscopy (FCS) compared to single molecule analysis is the possibility of averaging over many transits through the focal spot and thereby revealing typical characteristics of the molecules. However, all these methods are limited in their spatial resolution by diffraction. By applying the RESOLFT concept, with STED (STimulated Emission Depletion) the focal diameter can be seamlessly decreased from $\sim$250 nm to $\sim$40 nm allowing a much more precise detection in the spatial dimension (see figure 3.7). However, STED-FCS in three dimensions entailed a low signal to noise ratio and high background originating from insufficient fluorescence depletion in the axial out of focus areas (Ringemann, 2008).

Since the plasma membrane is a two-dimensional system, these problems were not pivotal for the investigation of the dynamic behavior and intermolecular interactions of phospholipids.

3.2.2 Incorporation of Lipids into the Plasma Membrane

Upon incorporation of the labeled lipids into the plasma membrane, interaction with the endogenous lipids is required. This is strongly dependent on the chemical structure of the lipids and also on their comparatively large fluorophore. The following lipids (see figure 3.8) were synthesized differing in their head group structure, length and number of the acyl chains, as well as type and labeling position of the fluorophore. The focus of investigation were two classes of lipids present in the plasma membrane, the phosphoglycerolipids (PE, PE-1 and DOPE) and the sphingolipids (SM, Cer and CPE as well as the gangliosides GM1, GM1-# and GM1-##).

These lipids are amphipathic, with their hydrophilic head group facing the aqueous phase (the cytoplasm and the extracellular environment, respectively) and their lipophilic long acyl chains repelling water, lining up against one another to form a bilayer with

3.2.2 Incorporation of Lipids into the Plasma Membrane

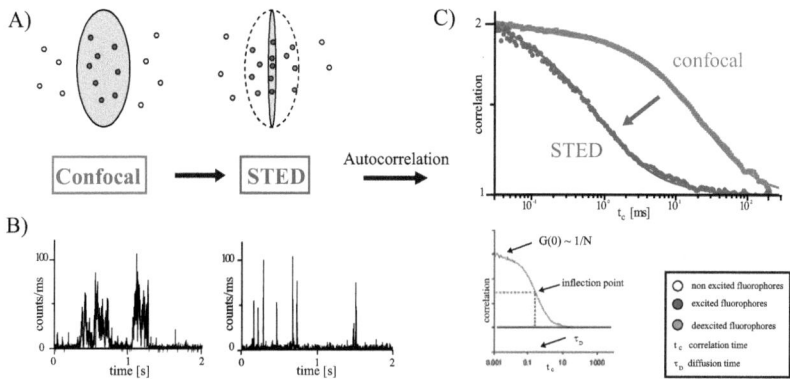

Figure 3.7: Enhancing of the spatial resolution by STED-FCS. Fluorescent molecules are diffusing through the confocal and STED focal volumes, respectively. In **A** the accordant effective focal volumes are shown (blue). In the confocal mode, many molecules contribute to the fluorescence signal, displayed in the left time trace in **B**. Upon application of the STED laser, only few molecules in the very center (at the zero node in the STED PSF) remain fluorescent. Therefore, the fluctuation in fluorescence intensity of only a few excited molecules is recorded, giving rise to very distinct time traces (**B**, right). After autocorrelation and normalization of these time traces, dynamic characteristics can be captured, e.g. the inflection point of the curve defines the diffusion constant τ_D (**C**). As an example the fluctuations of the fluorescence signal of a fluorescent labeled lipid (Atto647N-PE) were recorded in both, confocal and STED mode. The autocorrelated curves clearly show the shift to slower time constants for the STED (red) compared to the confocal case (green).

the heads on both sides facing the water. Because the labeled lipids spontaneously form micelles in an aqueous environment, incorporation into the plasma membrane of living cells is hindered.

The following subsection reports on two techniques which were applied here to incorporate the fluorescently labeled lipids into the plasma membrane of living PtK2 cells: via liposomes and via complex formation with BSA.

Incorporation of Labeled Lipids via Liposomes

For the introduction of labeled lipids via liposomes into the plasma membrane, a mixture containing both unlabeled and labeled lipids had to be prepared. For this purpose one type of labeled lipid (either PE-Atto647N, SM-Atto647N or DOPE-Atto647N) and a fixed set of unlabeled lipids (PE, PC, PS, PI and cholesterol) were mixed in a ratio resembling the

Results

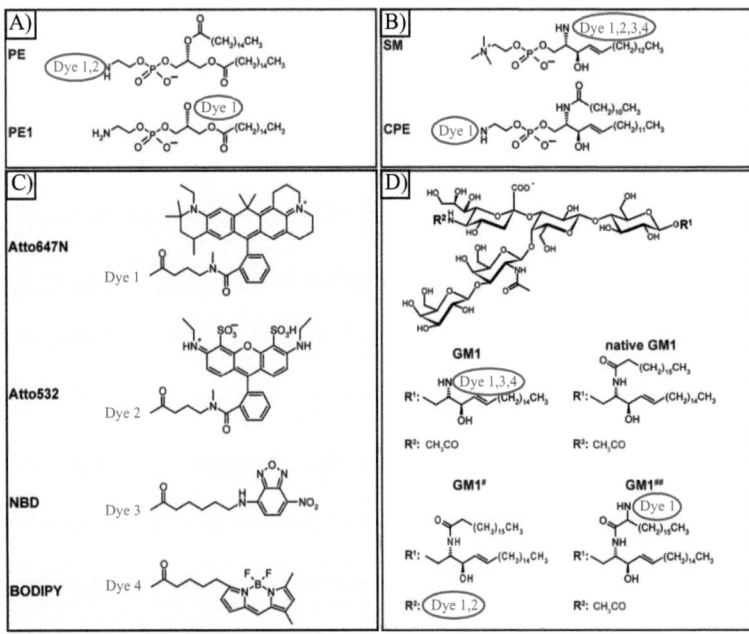

Figure 3.8: The structures of the used lipid analogs and dyes. In **A**, the structure of two phosphatidylethanolamine analogs (glycerol backbone), PE and PE1, is shown. Accordingly, the sphingolipids SM and CPE are shown in **B**. Note the different labeling positions (head or acyl chain replacement) in both the phosphoglycerolipids and the sphingolipids. The fluorescence excitation and emission maxima of the dyes highlighted in **C** are 645 nm and 670 nm, respectively for Atto647N (=Dye 1); for Atto532 (=Dye 2) 532 nm and 553 nm; for NBD (=Dye 3) 466 nm and 536 nm and for BODIPY (=Dye 4) 505 nm and 511 nm. The analogs of the mono-sialic ganglioside, GM1, native GM1, GM1-# and GM1-## are listed in **D**. The red numbers in the lipid structures indicate which dyes were used for conjugation.

endogenous lipid composition. In aqueous solution this lipid mix rapidly assembled into bilayer vesicles, liposomes, which were then extruded through a polycarbonate membrane (see in detail in subsection 5.2.1 and table 5.3), resulting in liposomes with an average diameter between 50 nm to 80 nm.

The fusion of the liposomes with the plasma membrane of living cells was performed under distinct environmental conditions which were obtained by using a fusion buffer with high calcium concentrations (see table 5.1). After fusion a brief ultrasonic pulse was

3.2.2 Incorporation of Lipids into the Plasma Membrane

applied to remove the upper plasma membrane. Thus the background emerging from liposomes sticking on the plasma membranes could be suppressed. These so called membrane sheets were kept in a glutamate buffered system mimicking the cytosol to ensure proper function. However, less than 50% of all cells were transformed into sheets after that procedure, as could be verified by TMA-DPH staining (UV excitation). This lipophilic dye shows strongly enhanced fluorescence upon incorporation into a lipid bilayer. The sheets could be identified by a 50% weaker fluorescence compared to intact cells because only one membrane, the one facing the coverglass, was stained (see also subsection 5.5.1).

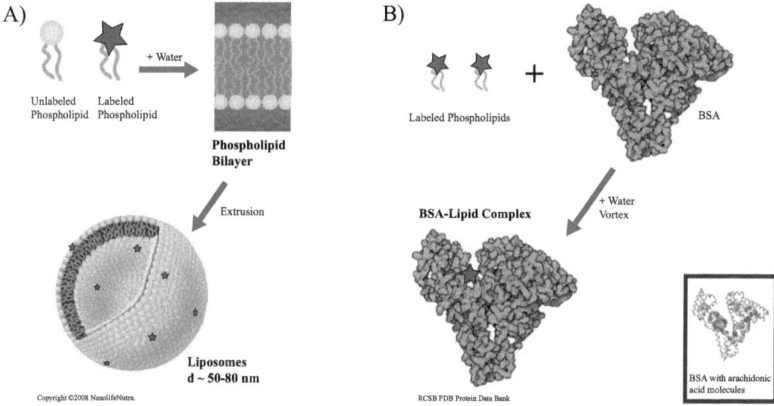

Figure 3.9: Generation of liposomes versus BSA-lipid complex formation. The generation of liposomes (shown in **A**) is done by drying unlabeled phospholipids (PC, PE, PI, PS) and cholesterol together with one labeled lipid (e.g. Atto-647N-PE) under a stream of liquid nitrogen. Upon swelling in aqueous solutions the lipids spontaneously form bilayers. Unilamellar liposomes with a diameter of $d \sim 50 - 80$ nm are generated by 50 cycles of extrusion through a polycarbonate membrane.

However, for the measurements of lipid dynamics using STED-FCS, TMA-DPH was not introduced, both to avoid possible background fluorescence and severe changes in the membrane properties due to an excess of TMA-DPH labeling and the harsh UV excitation. Thus, a second control sample was prepared in parallel for wide field fluorescence microscopy (refer to subsection 5.6.1) for each experiment.

To control the temperature in the sample, the cells were mounted in a special microscope chamber with an objective heater. The control sample was also mounted in a microscope chamber, though without temperature control, and was checked for the quality

Results

of the sheet generation by TMA-DPH staining. After the sheet generation the cells were exclusively kept in a calcium-free sonication buffer to avoid triggering off exocytosis.

However, gathering statistically relevant data for the STED-FCS dynamic measurements turned out to be challenging. The time window (about ten minutes) was simply too small to find the plasma membrane sheets through a 100 x objective lens, before severe changes in the membrane sheets occured. Another disadvantage of this incorporation technique was the non uniform labeling efficiency throughout the sample due to inhomogeneous fusion events in individual cells.

Thus, another approached was employed as described in the following subsection.

Incorporation of Labeled Lipids via BSA-coupling

A different method, the complex formation of lipids with BSA made the use of liposomes and membrane sheets dispensable. This technique was established by Pagano and Martin (Pagano et al., 1989) and later modified by Günter Schwarzmann (private communication, refer to subsection 5.2.3). The difference between Pagano and Martin's approach and Günter Schwarzmann's method was the amount of ethanol used for redissolving the nitrogen dried lipids (Pagano: 200 μl resulting in a final ethanol concentration of 2% (v/v) vs. Schwarzmann: 20 μl leading to 0.2% (v/v)) and the amount of defatted BSA (0.17 mg/ ml and 0.7 mg/ ml, i.e. in a total volume of 10 ml $\sim$ 25 nmol and 100 nmol, respectively). Using only 20 μl ethanol, no dialysis was required since the resulting concentration is tolerated by the cells. Moreover, the incorporation rate of the labeled lipids was higher for Schwarzmann's technique, possibly due to the higher BSA concentration. However, the remaining problem was to introduce larger more complex lipids in the membrane in satisfactory amounts.

Therefore, this technique was further modified in this thesis to finally introduce the larger diacyl-phosphoglycerolipids into the plasma membrane. This was accomplished by increasing the final BSA amount to 200 nmol instead of 100 nmol, leading to a lipid to BSA ratio of 1:2 (see figure 3.10). Having a closer look at the acyl chain to BSA ratio a 1:1 dependency was found, as it was the case for the monoacyl-phospholipids (A detailed discussion is found in subsection 4.2.3). For all the following lipid incorporations, the BSA concentration was chosen to be either 0.7 mg/ ml or 1.4 mg/ml, corresponding to 100 nmol and 200 nmol of BSA, respectively, depending on the number of native acyl

3.2.2 Incorporation of Lipids into the Plasma Membrane

chains. The incubation time on ice for both mono- and diacyl lipid analogs was kept constant at 30 min.

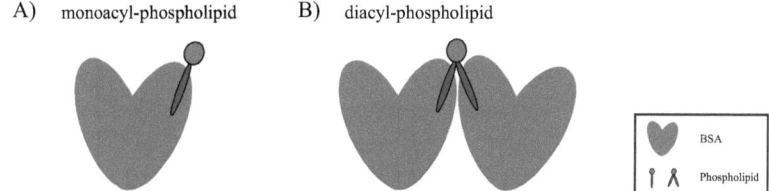

Figure 3.10: Model of BSA complexing diacyl-phospholipids. A The single fatty acid residue of monoacyl-phospholipids resides in one of the hydrophobic pockets of the BSA molecule, leaving the hydrophilic head facing the aqueous environment. **B** On the other hand, the two fatty acids of diacyl-lipids are complexed by two BSA molecules in this model.

Some lipids (PE, PE-1, DOPE, SM and CPE) required an additional incubation step at 37 °C for 4 min to obtain a homogeneous distribution in the plasma membrane. This additional incubation time was determined by balancing out an even distribution of the lipids in the plasma membrane and an undesired increase of internalization events, both scaling with time. All labeled lipids used for BSA conjugation are listed in table 5.4. The overall amount of incorporated lipids could be controlled easily by adjusting the final concentration of the lipids used for incubation, thus enabling concentrations convenient for single molecule measurements ($\sim$nM) up to imaging ($\sim \mu$M). The not incorporated lipids could be easily washed away so that background fluorescence was negligible.

Thus all measurements were performed on the lower membrane of intact cells without the need of sheet generation. Moreover, all dynamic measurements were completed before significant internalization of the labeled lipids or severe morphological changes in the cells could occur. Most dynamic measurements were done at 27 °C or 37 °C. For discussion about the temperature dependence of the lipid's diffusion please refer to subsection 4.2.9.

Results

3.2.3 The Influence of the Fluorescent Label on the Lipid Behavior

By using fluorescent dyes in fluorescence microscopy one risks to alter the properties of the labeled molecule. Since the molecular weight of the dyes (∼0.6 kDa) is in the same range as the molecular weight of the used lipids (∼0.7-1.5 kDa), the influence of the label, its polarity or size, can be crucial. As already mentioned, the studied lipids are amphipathic. By introducing labels with different polarities this amphipathy could be changed, resulting in an improper anchoring of the labeled lipid into the plasma membrane. As shown before in a model membrane system (Wang and Silvius, 2000), the replacement of one of the acyl chains of a sphingolipid by a short linker acyl chain carrying a dye molecule (NBD or BODIPY) led to the tendency to partitional loss in sphingolipid-enriched domains than sphingolipid analogs, whose acyl chains resemble more those of native sphingolipids. Therefore, various control experiments on several fluorescent (and unlabeled) lipid analogs, differing in their structure, in the labeling position, and in the polarity of the dye were performed. The following subsections report on these in detail.

Summing up the control experiments, the lipophilic dye Atto647N performed best in comparison with the very hydrophilic dye Atto532, and the commonly used hydrophilic dyes NBD and BODIPY FL. All Atto647N-labeled lipids showed no observable influence on the lipids. Depending on the labeling position, that was also true for some of the Atto532-labeled lipids. The NBD- and BODIPY-labeled lipids, however, although sterically smaller, significantly influenced the lipid's behavior.

Investigation of the Membrane Affinity of the Labeled Lipid Analogs

The membrane affinity of the fluorescent lipid analogs was checked by back exchange experiments. These experiments were done with confocal laser scanning microscopy (see subsection 5.6.2). The cells were prepared in the same manner as for STED-FCS measurements. For control images, the labeled cells were mounted in cold HDMEM ($\sim 4\,°C$) to avoid rapid internalization. The back exchange was performed at $4\,°C$ with 10^{-5} M defatted BSA (for the protocol in detail refer to 5.2.3). By comparing the fluorescence signal of the images taken before and after washing, the removal of the fluorescent lipid analogs from the plasma membrane could be determined. For proper comparison of the lipid analogs of SM and GM1, respectively, the same labeling position, namely at the

3.2.3 The Influence of the Fluorescent Label on the Lipid Behavior

water-lipid interface, was used here.

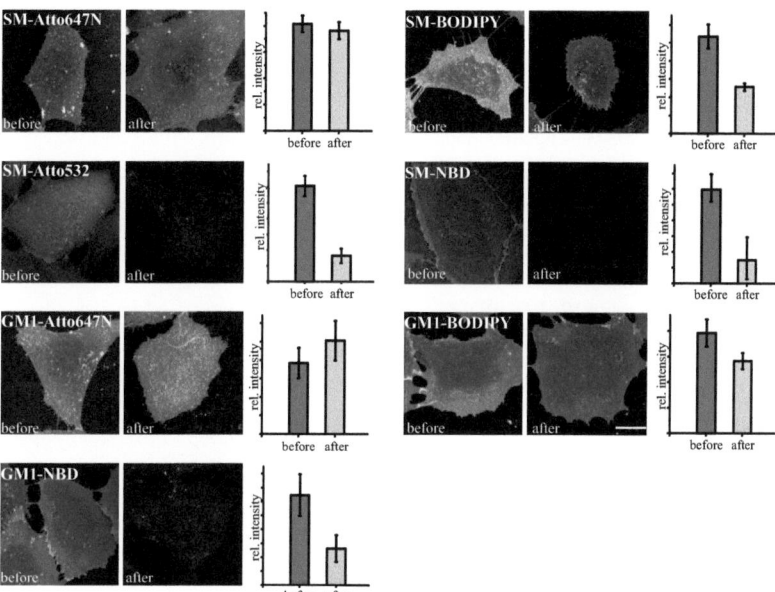

Figure 3.11: Verification of lipid incorporation by BSA-depletion. Scanning confocal images of the plasma membrane of living PtK2 cells incorporating Atto647N-, Atto532-, NBD- or BODIPY-labeled SM or Atto647N-, BODIPY- or NBD-labeled GM1 before (left) and after (right) the ten-minute washing with 10^{-5} M BSA. The column diagrams show the averaged fluorescence signals taken from several cells also before and after back exchange. Note the internalization of the lipids after a certain time, resulting in both bright, clustered spots inside the cells in the confocal images and a slight increase in the averaged fluorescence signal of SM-Atto647N. The settings for taking the images before and after the BSA depletion were kept constant. The scale bar applies for all images: 20 µm

The results (shown in figure 3.11) clearly revealed distinct dissimilarities of the incorporation efficiencies of the labeled SM and GM1 analogs depending on the fluorescent labels. The SM analogs labeled with the three hydrophilic dyes BODIPY, Atto532 and NBD showed a residual fluorescence signal after BSA washing of approximately 48%, 27% and 25%, respectively, whereas the SM-Atto647N analog retained a fluorescence intensity of 94%. This strongly argued for the correct anchoring of the labeled lipid analog in the plasma membrane in the case of SM-Atto647N unlike for the three other

Results

hydrophilic dyes. The tendency for the fluorescently labeled GM1 analogs is similar but less pronounced. The increase of the fluorescence signal after BSA washing for GM1-Atto647N was due to the rapid internalization (stained vesicles) resulting in a higher intracellular background (note that the background on the coverslip did not increase, arguing for internalization against free floating lipids). This higher cellular background could also be observed in the GM1-BODIPY and in the GM1-NBD samples. In the latter case besides the fluorescence stemming from the vesicles almost no plasma membrane staining was observable.

The conclusion drawn from these experiments was that the proper anchoring of the labeled lipids strongly depends on the type of dye. The short acyl chain carrying the hydrophilic dyes Atto532, BODIPY or NBD could not been integrated into the plasma membrane, therefore facilitating the back exchange of the lipid by improper anchoring. On the other hand, the short acyl chain carrying the lipophilic dye Atto647N showed the same properties as the lipophilic membrane anchor, thus hampering the back exchange of the fluorescent lipid analog by BSA.

Investigation of the Lateral Diffusion of Labeled Lipids using FRAP

Since Atto647N-labeled lipids performed best in the back exchange experiments, they were further investigated for their lateral diffusion ability within the plasma membrane by using FRAP (Fluorescence Recovery After Photobleaching). Since in FCS experiments only low concentrations of the labeled molecules are used, a control is required to analyze whether the observed diffusion is due to lateral diffusion within the membrane. A detailed protocol for FRAP experiments is given in subsection 5.2.3.

PE and SM, one glycerophospholipid labeled in the water phase and one sphingolipid labeled in the water-lipid interphase, were incorporated in the plasma membrane of living PtK2 cells. Then a small area ($\sim 15 \times 15 \,\mu m^2$) was rapidly photobleached (see figure 3.12 **B**) by applying high laser power. Within this area the recovery of the intensity was monitored. If lateral diffusion is feasible, the photobleached dyes diffuse out and new fluorophores diffuse in. By knowing the diffusion length and the time, the diffusion constant D of the labeled lipids can be obtained with:

$$D = \frac{r^2}{4t_{1/2}}$$

3.2.3 The Influence of the Fluorescent Label on the Lipid Behavior

With r defining the radius of the bleached area and $t_{1/2}$ describing the time when 50% of the fluorescence had recovered. The diffusion constant for PE was $D_{PE} = (5.5 \pm 1.0) \times 10^{-9}\,\text{cm}^2/\text{s}$ and for SM $D_{SM} = (5.5 \pm 1.0) \times 10^{-9}\,\text{cm}^2/\text{s}$, which is comparable to the diffusion constant obtained with STED-FCS (Ringemann, 2008).

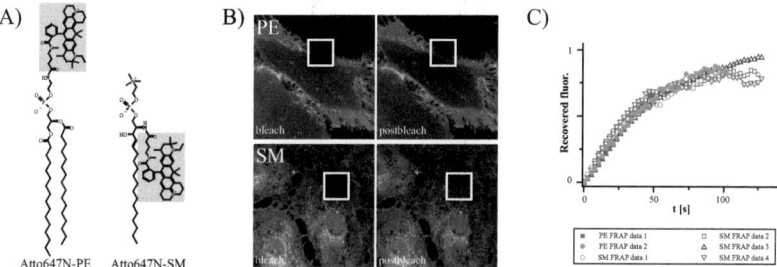

Figure 3.12: FRAP experiments of Atto647N-labeled SM and PE. In **A** the molecular structure of both lipid analogs is shown, with the dye highlighted in gray. PtK2 cells, incubated with either SM or PE, were imaged using standard confocal laser microscopy. Then a small area of the plasma membrane was photobleached (shown in **B**) and the recovery of the fluorescence monitored while bleached lipids diffuse out and new lipids in.

Determination of the Lipophilicity of the Labeled Lipid Analogs

By analytical high performance thin layer chromatography (HPTLC) of the labeled and native ganglioside analogs (as shown in figure 3.13; for structural details please refer to figure 3.8), the lipophilic character and its dye induced changes were analyzed. Native GM1, Lyso-GM1 (unlabeled, with only one long acyl chain), NBD-GM1 and Atto647N-GM1 (label replaces the long acyl chain by a short acyl chain carrying either NBD or Atto647N), Atto647N-GM1-## (dye anchor in addition to the two native acyl chains), and Atto647N-GM1-# and Atto532-GM1-# (labeling at the polar carbohydrate domain of the head group) were analyzed.

The HPTLC (see figure 3.13) showed that the chromatographic behavior of Atto647N-GM1 (acyl chain replacement) or Atto532-GM1-# (head group) resembled that of the native GM1, leading to the assumption of very similar lipophilic characteristics. Atto647N-GM1 (acyl chain replacement) was even slightly more lipophilic than its native counterpart unlike NBD-GM1 (acyl chain replacement) which was slightly less lipophilic.

Results

Atto647N-GM1-## (additional acyl chain), however, was even more lipophilic. For proper incorporation into the plasma membrane, the lipids require strong anchoring, which is provided by two long acyl chains in the case of native GM1. The most pronounced hydrophilicity could be observed for the unlabeled Lyso-GM1, since it possesses only one acyl chain.

The labeling of the polar carbohydrate domain of GM1 with Atto647N and Atto532, respectively, led in the case of Atto647N to a shielding of the head group's polarity, resulting in an overall increase of the lipophilic character of Atto647N-GM1-# in comparison to Atto647N-GM1 and -GM1-## where the label was attached to a non polar acyl chain. From this and the previous back exchange results, one could expect that the Atto647N-carrying acyl chain is embedded within the lipid layer similar to the native GM1, thus resulting in membrane anchoring.

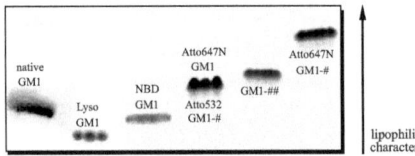

Figure 3.13: Chromatographic behavior of the different ganglioside analogs. HPTLC of native and labeled GM1 analogs (SiO$_2$): native, NBD- and Atto647N-GM1 (label at the water-lipid interface), unlabeled Lyso-GM1, and Atto647N-labeled GM1-# (label at the head group) and GM1-##(at the water-lipid interface). Note the chromatographic behavior of Atto532-GM1-#, two native acyl chains, is the same than for Atto647N-GM1, one acyl chain native, the other one carrying the dye. The Atto647N-labeled GM1 analogs are comparable or slightly more lipophilic than their native counterpart, whereas on the contrary, NBD-GM1 is slightly less lipophilic. The GM1 analog lacking one acyl chain, Lyso-GM1, showed the most pronounced hydrophilicity. The attachment of Atto647N to the carbohydrate head group of GM1, as it is true for GM1-##, resulted in the highest observed lipophilicity. The polar head group is shielded by the dye and as such increasing the overall lipophilicity.

3.2.3 The Influence of the Fluorescent Label on the Lipid Behavior

Determination of the Lipid Properties Dependence on the Labeling Position

To investigate the effect of the dye in respect to the labeling position, both lipid classes, the phosphoglycero-(PE) and sphingolipids (SM and GM1), were labeled with Atto647N at different positions: in the water phase (at the head group) or at the water-lipid interface (the native long acyl chain was replaced by a short acyl chain carrying the dye). The two classes differed in their diffusion characteristics but were very similar within one class, irrespective of the labeling position (see figure 3.14). The free diffusion coefficient determined by FCS analysis was the same for both PE analogs, Atto647N-PE (head group) and Atto647N-PE-1 (acyl chain replacement). The unsaturated PE analog DOPE, featuring a kink in both acyl chains due to the double bonds, showed the same diffusion characteristic than its saturated counterpart, although it is sterically larger.

Also the trapping characteristics of Atto647N-CPE or Atto647N-GM1-# (head group) were unchanged compared to Atto647N-SM, Cer-Atto647N and Atto647N-GM1 (acyl chain replacement), respectively. Since CPE and SM only differed in their acyl chain structure, the trapping phenomenon was possibly only evoked by the ceramide backbone itself (sphingosine linked via an amide to a fatty acid). As shown in figure 3.14 **B**, the ceramide analog Cer-Atto647N showed trapping as well which could be abolished upon COase addition.

In control experiments, where cells were incubated with the free dye Atto647N no trapping was observed. Therefore it was concluded that the trapping observed for the Atto647N-labeled ceramide, sphingolipids and gangliosides was not caused by interactions of the dye with other membrane components.

By replacing the lipophilic dye Atto647N with the hydrophilic dye Atto532, a general change in the lipid's behavior could be observed: The anomaly for Atto532-SM (label at the water-lipid interface) still remained high with $1/\alpha > 1.5$ (figure 3.14 **C**). On the other hand the diffusion time was significantly shorter and no trapping could be detected, leading to the conclusion that Atto532-SM could not interact with other membrane components. Also the back exchange results (see figure 3.11) confirm for Atto532-SM an improper incorporation into the membrane. Since in both Atto647N- and Atto532-labeled SM, the identical sphingolipid was used, one could conclude that the change in behavior was due to the dye. However, Atto532-PE and -GM1-#, both labeled in the water phase, showed no change in diffusion compared to the accordant Atto647N-lipids (figure 3.14

Results

B). This can be explained by the polarity of the huge head group itself, shielding possible influences of the dyes.

In conclusion, taking all previous experiments (back exchange, TLC, FCS) into account, no change in the diffusion characteristics was observable upon attachment of the lipophilic dye Atto647N, irrespective of the labeling position. For the hydrophilic dye Atto532, that was also true if the label was placed on the polar head domain, in the water phase, as it was the case e.g. for Atto532-GM1-#. As expected, proper membrane anchoring was hampered when replacing the acyl chain with a hydrophilic dye (Atto532) whereas both dyes, hydrophilic and lipophilic, showed no change on the lipid's diffusion when attached to the polar head group. For this reason, Atto647N was the most suitable label for studying the lipid dynamics.

3.2.4 Measurements of Lipid Dynamics in the Plasma Membrane

Atto647N-labeled phosphatidyl ethanolamine analogs (PE, glycerol backbone) and sphingomyelin (SM, ceramide backbone) were incorporated into the plasma membrane. For simplicity, if Atto647N is used as label, the labeled lipids are only referred to as e.g. PE or SM. Time traces of the fluorescent molecules were recorded in both modes, confocal (d $\approx$ 250 nm) and STED (d $\approx$ 40 nm). After correlation and normalization of these traces, the resulting curves were fitted with mathematic models to describe the molecule diffusion.

SM unlike PE is presumably integrating into microdomains smaller than 200 nm (see subsection 3.2.5). In the analysis, it was assumed that the brightness is equal for free and hindered transit events (Ringemann, 2008) with amplitudes A_1 and A_2 and transit times τ_{D1} and τ_{D2}.

Assuming that $A_1 + A_2 = 1$, the amplitudes A_1 and A_2 directly represent the fraction of free and hindered molecules. The diffusion time τ_{D1} was chosen to denote the free diffusion that depends linearly on the size of the detection area (with a diffusion coefficient derived from the measurements of the freely diffusing PE), whereas τ_{D2} was a free parameter describing the hindrance or trapping of the molecules.

For the anomalous diffusion model a new parameter $1/\alpha$ was introduced describing the degree of hindrance. High values for $1/\alpha$ account for a higher hindrance, whereas a

3.2.4 Measurements of Lipid Dynamics in the Plasma Membrane

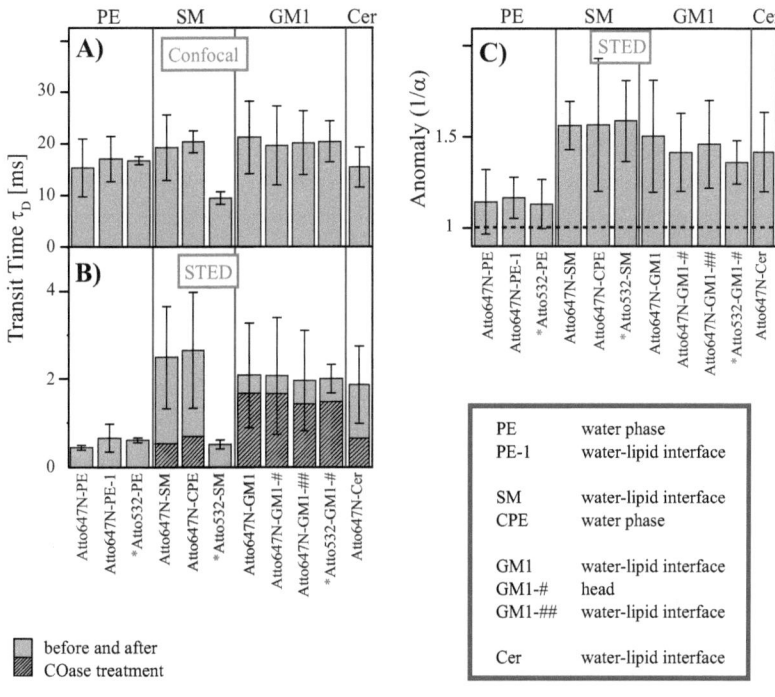

Figure 3.14: The influence of the labeling position on the lipid's dynamic behavior. FCS analysis applying the model of anomalous diffusion for **A** confocal (d≈250 nm) and **B, C** STED (d≈40 nm). The average transit time τ_D of Atto647N-PE, -PE-1, -SM, -CPE, -Cer, -GM1, -GM1-#, GM1-## and Atto532-PE, -SM, -GM1, respectively, is shown in **A.** and **B.**, whereas their anomaly (1/α) is shown in **C.**. The blue box highlights the labeling sites of the lipids. The transit times for ceramide, the sphingolipids and gangliosides after cholesterol depletion by COase are displayed by the meshed bars in **B**. Note the diffusion and trapping behavior of Atto647N-Cer strongly resembles the one of the sphingomyelins. Since another STED setup was used for measuring the Atto532-labeled lipids (indicated by the red asterisks) the focal spot was slightly different (d≈200 nm for confocal and d≈60-70 nm for STED) due to the different laser wavelengths used and STED intensities applied. The transit times were extrapolated to the values expected for $d \approx 250$ nm and $d \approx 40$ nm respectively. The error bars resulted from averaging over more than 30 measurements. This figure is published (slightly altered) in (Eggeling et al., 2009).

value of 1 corresponds to a free diffusion.

As shown in figure 3.15 **A** in the confocal mode (d ≈250 nm) both lipids, PE (red

Results

dots) and SM (gray dots), could be perfectly described using a model featuring only one transit time (red lines). The transition times in confocal mode obtained for PE and SM were $\tau_D(\text{PE})$ =19 ms and $\tau_D(\text{SM})$ =28 ms, respectively. The confocal recordings could not explain whether the slightly prolonged focal transit time of SM was due to transient local trapping or due to a slower diffusion constant. However, as depicted in figure 3.15 **B**, PE recorded in the STED-FCS mode was significantly shifted to shorter correlation times, $\tau_D = 0.45$ ms. In accordance with the confocal recordings, the PE data could be displayed with a single-species fit. The correlation curve for SM, in contrast, had to be described using two dissimilar modalities or, alternatively, anomalous diffusion. By that the focal transition of SM can be described assuming both, a free diffusion with $\tau_{D1} = 0.45$ ms and a hindered diffusion defined by $\tau_{D2} = 10$ ms and $A_2 = 64\%$, giving the fraction of molecules that are hindered or trapped.

Two different analogs of PE were investigated, differing in their labeling positions: label at the head domain (PE) versus acyl chain replacement (PE1). Furthermore, also the diffusion behavior of the unsaturated diacyl analog DOPE, consisting of two mono-unsaturated acyl chains, resembled the diffusion characteristics of the saturated analogs PE and PE1 and was also independent of cholesterol (data not shown). As these analogs showed no difference in their diffusion behavior, the figure simply refers to them as PE.

3.2.5 Influence of Cholesterol on the Diffusion of SM, GM1 and Cer

The investigation of the presumptive cholesterol dependent trapping of sphingolipids was carried out depleting the endogenous cholesterol, either oxidizing cholesterol enzymatically by cholesterol oxidase (COase) or by its removal from the plasma membrane with β-cyclodextrin (β-CD). The oxidoreductase COase catalyzes the reaction between cholesterol and oxygen to cholestenone (cholest-4-en-3-one) and H_2O_2 with an approximated efficiency of $\sim$80% (Pörn and Slotte, 1995). The oligosaccharide β-CD, on the other hand, lodges the bulky hydrophobic cholesterol inside its ring structure, as such removing it from the lipid bilayer with an efficiency of $\sim$60-90% (Pike and Miller, 1998; Zidovetzki and Levitan, 2007; Castagne et al., 2008).

Both treatments were performed and compared to compensate for eventual negative side effects caused by the individual reagent (e.g. H_2O_2 generation by COase). In prin-

3.2.5 Influence of Cholesterol on the Diffusion of SM, GM1 and Cer

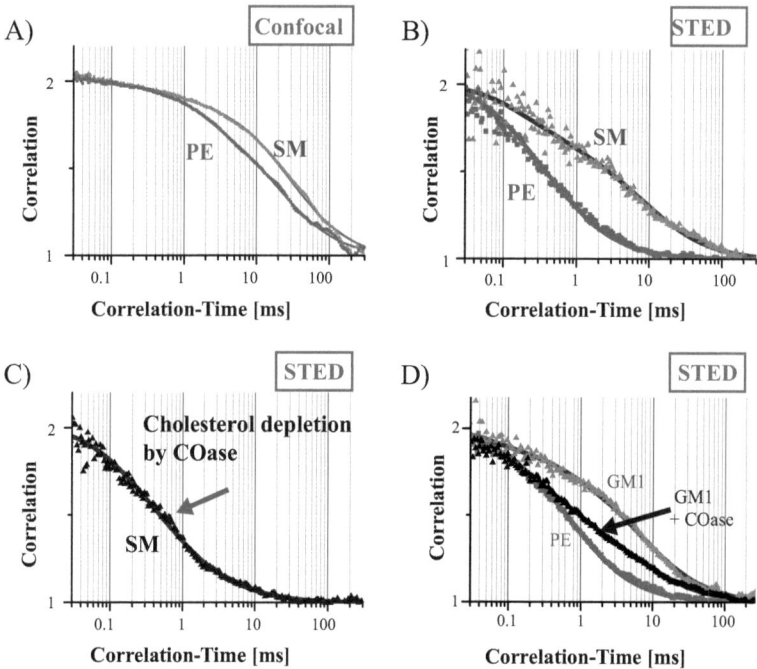

Figure 3.15: Fluorescence Correlation Spectroscopy of PE and SM in the plasma membrane. A Normalized confocal correlation data of Atto647N-PE (red dots) and Atto647N-SM (gray dots). In the confocal case (d ≈ 250 nm) both curves could be fitted with a single-species fit (red lines) with transit times of $\tau_D = 19$ ms and 28 ms, respectively. The confocal data could not reveal whether the slightly prolonged focal transit time of SM was due to transient trapping or just due to slower diffusion. **B** Corresponding normalized STED correlation data (d ≈40 nm) of PE and SM. In comparison to the confocal recordings, the PE correlation curve was significantly shifted to shorter correlation times following the reduction in detection area. The SM data gathered with the STED recordings revealed, however, that the shift to longer time scales was due to hindered diffusion. The PE-data could be fitted with the red line, a single-species fit ($\tau_D = 0.45$ ms), whereas the blue line fit of the SM recordings assumed two dissimilar modalities of focal transits, namely a free and a hindered diffusion ($\tau_{D1} = 0.45$ ms, $\tau_{D2} = 10$ ms and $A_2 = 64\%$, the fraction of hindered molecules). **C** Normalized STED correlation data of SM diffusion after cholesterol depletion by COase (black dots). With decreased levels of cholesterol in the plasma membrane SM now diffused similar to PE. Note the shift to shorter time scales indicated by the red arrow. **D** Normalized STED correlation data (d ≈40 nm) of GM1 (gray dots), GM1 upon COase addition (black dots) and for comparison PE (red dots) are shown. This figure is published (slightly altered) in (Eggeling et al., 2009).

Results

ciple it was irrelevant whether cholesterol depletion was carried out before or after the fluorescent lipid incubation. However, if the depletion was performed after the incorporation with the labeled lipids, internalization of the fluorescent lipids was observed, leading to an increased background. Therefore, for most of the recordings, cholesterol depletion was carried out before the incubation with the labeled lipids. The depletion was reversible for both reagents in the absence of COase or β-CD, respectively. Not surprisingly, the recovery of cholesterol was slower at 27 °C compared to 37 °C, leaving a longer time window for the recordings at the lower temperature. Every cell was judged by its appearance in transmission light for its viability directly before recording.

Upon cholesterol depletion by COase, the correlation curve for SM was comparable to the one obtained for the freely diffusing PE (as depicted in figure 3.15 **C**). PE was unaltered by the COase treatment, arguing against artifacts introduced by the cholesterol conversion. The average diffusion time τ_D of SM showed a significant shortening due to the increased fraction of freely diffusing molecules. However, its diffusion behavior maintained anomalous with a remaining fraction of trapping events ($\tau_{D1} = 0.45$ ms, $\tau_{D2} = 4$ ms and $A_2 = 15\%$). These results may be due to incomplete depletion by COase efficiency to convert cholesterol accompanied by the recovery of the endogenous cholesterol content. Upon cholesterol depletion with β-CD, the diffusion of both lipids was similar compared to that after the COase treatment, although no anomaly in the SM diffusion could be observed. This controversy remained speculative and is further discussed in subsection 4.2.7. The different impacts of the treatments on the F-actin cytoskeleton are discussed in the following subsection.

Not surprisingly, another type of sphingolipid, the glycosylated ganglioside GM1, also showed a hindered diffusion behavior similar to SM. Similar to the SM recordings in **B**, two modalities of focal transits had to be assumed for GM1 ($\tau_{D1} = 0.5$ ms, $\tau_{D2} = 8.5$ ms and $A_2 = 45\%$), fitted with the blue line, whereas the PE data were represented by a single-species fit (red line) with $\tau_D = 0.5$ ms. Gangliosides and sphingomyelin share the ceramide backbone (sphingosine + fatty acid) which is embedded in the lipid phase of the membrane. They only differ in their head group structure, namely a phosphocholine for SM, and a sialic acid linked to an oligosaccharide for GM1, respectively. Therefore, upon COase adddition, trapping was partially abolished and the fraction of hindered diffusing GM1 molecules was only slightly reduced to 30% due to cholesterol depletion ($\tau_{D1} = 0.5$ ms, $\tau_{D2} = 9$ ms and $A_2 = 30\%$). The diffusion behavior strongly resembled the one of SM, although the fraction of trapped molecules was smaller and the effect of COase depletion less pronounced (compare with figure 3.15 **D**).

Since cholesterol interaction of the sphingolipids is possibly established by their backbone (ceramide), an Atto647N-labeled ceramide (Cer) was incorporated into the plasma membrane and its diffusion behavior investigated. Trapping similar to SM and GM1 was also detected for Cer and cholesterol dependency was verified by a decreased transit time upon depletion of cholesterol with COase (see figure 3.14 **B**). In accordance with the other lipids, ceramide also exhibited anomalous diffusion as depicted in figure 3.14 **C**.

In conclusion, cholesterol plays an important role in the trapping of the sphingolipids by possibly interacting with the ceramide backbone. This interaction could be abolished by removal or oxidation of cholesterol. The head domains of the sphingolipids had little (gangliosides) or no effect (sphingomyelin) on the trapping, as ceramide alone showed the same behavior.

3.2.6 Effectivity of Cholesterol Depletion

The actin cytoskeleton is linked via spectrin and other actin-binding proteins to the plasma membrane. The loss of integrity of lipid microdomains upon the depletion of cholesterol from the plasma membrane also has downstream effects on the cell cortex and the actin cytoskeleton in particular (Qi et al., 2009). Therefore, by investigating the effects of cholesterol depletion on the F-actin cytoskeleton, conclusions can be drawn about the efficiency of the depletion.

PtK2 cells were treated as described above with either COase or β-CD, directly followed by formaldehyde fixation. F-Actin was visualized with a rhodamine-phalloidin staining and recorded with a standard confocal laser scanning microscope. Upon both treatments, a reorganization of F-actin stress fibers is observable. Usually, PtK2 cultures form closely lined up monolayers. They are laterally connected by tight junctions. A prominent feature of these cultures is their actin cytoskeleton, since it appears as distinct stress fibers (see figure 3.16, control).

In COase treated cells, the stress fibers alongside their periphery are strengthened (white arrows) and the cell borders are more prominent. In accordance with Qi *et al.* (Qi et al., 2009) stress fiber strengthening is an indicator of a decrease in the cholesterol

Results

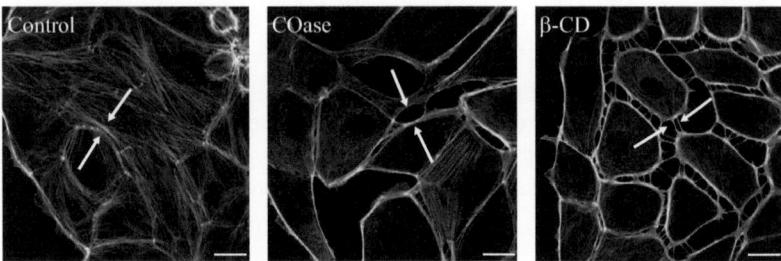

Figure 3.16: The Effect of Cholesterol Depletion on the Actin Cytoskeleton. The endogenous cholesterol content of PtK2 cells was depleted by either incubation with the oxidoreductase cholesterol oxidase (COase) or by application of the oligosaccharide β-Cyclodextrin (β-CD). After this treatment cells were fixed followed by phalloidin-rhodamine staining to visualize the F-actin filaments. The white arrows indicate the cell borders. All recordings were performed using exactly the same settings (laser intensity, detector gain, image dimensions). The length of the scale bar is 20 μm.

level. Therefore it was concluded that also in the lipid diffusion experiments the COase treatment led to decrease in the endogenous cholesterol level of the plasma membrane.

The treatment with β-CD had a more prominent effect on the F-actin cytoskeleton and furthermore on the cell morphology itself. The cells do not line up alongside their periphery, but are only connected by their intercellular junctions which are indirectly visualized by the strong phalloidin staining.

Independent from their different impacts, both treatments were reversible after recovery in normal culture medium using standard culture conditions. It was concluded that both treatments led to a sufficient cholesterol depletion albeit to a different extent.

3.2.7 Approximation of the Spatial Dimensions of Microdomains

According to Ringemann, the length scale in which trapping of SM takes place could be approximated using the dependence of the transition time τ_D and the fraction of trapped molecules A_2 on the focal diameter as a hint (Ringemann, 2008). As already mentioned earlier, the focal spot of the STED-FCS setup could be continuously varied between $\sim$250 nm and $\sim$30 nm. At a focal diameter of 160 nm the heterogeneous diffusion of SM became apparent, and it was significant for focal diameters below 60 nm (Ringemann,

3.2.7 Approximation of the Spatial Dimensions of Microdomains

2008). As expected for free diffusion, the transition time for PE scaled linearly with the focal diameter. For the fraction of hindered SM molecules A_2, a plateau was reached at $\tau_{D2} \sim 10$ ms for focal diameters of 50 nm and below. This was independent of τ_{D1}, therefore one could assume that only one trapping event occurred on average. Assuming that the molecule's passage consists of two parts, a free diffusion part and eventual trapping events, the actual trapping time could be approximated with $\tau_{trap} = \tau_{D2} - \tau_{D1}$, thus simply subtracting the free diffusion time τ_{D1} from the hindered part τ_{D2}.

Results

4 Discussion and Outlook

4.1 Post-translational Modifications of Neurofilaments

Post-translational modifications (PTMs) are functional highly important since they can alter the chemical properties of the protein by e.g. modifying its amino acids by attaching functional groups, removing amino acids (such as methionine) or changing its structure by the formation of disulfide bonds. Many PTMs are highly dynamic. This allows the cell to adapt to changes quickly. The neurofilament triplet proteins are mainly modified by phosphate groups or glucosamine, which influence the assembly and disassembly of the subunits. Not only the kind of PTM, but also its position on the NF protein is an important parameter that determines the effect.

The investigation of post-translationally modified neurofilament subunits requires besides affinity markers such as antibodies, also a suitable model system. Here, a human neuroblastoma model was used, considering that PTMs and their regulation also vary between different organisms. These model systems can be redifferentiated to obtain postmitotic neuron-like cells. The first step in this redifferentiation was the sensitization of the cells for neurotrophic factors upon the application of retinoic acid, a retinoid. This and further effects of retinoids on cells, and in particular on cancer cells, are discussed below.

4.1.1 The Effects of Retinoids on Cancer Cells

Various retinoids have been shown to modulate growth and differentiation in a number of human cancers, which are reviewed in (Gudas, 1994). They further play a crucial role in the clinical treatment of neuroblastoma, the most abundant solid childhood tumor with a

Discussion and Outlook

very low survival rate (Wagner and Danks, 2009), since retinoids, such as 13-*cis* retinoic acid, exert growth inhibitory and differentiation effects. These are mediated by nuclear retinoic acid receptors (RARs) or retinoic X receptors (RXRs). They normally function as ligand-inducible transcription factors by binding to response elements (RAREs) that are located in the regulatory region of target genes (Carpentier et al., 1997). Furthermore, due to their redox properties, they also enhance immune response, gap junctional communication, carcinogen-metabolizing enzyme activity and modulation of the intracellular redox status (Palozza et al., 2006).

SH-SY5Y cells, a clonal derivative of the human neuroblastoma SK-N-SH cell line, expresses high levels of RAR (Lovat et al., 1993), thus making them sensitive to retinoids. The retinoid used here to induce differentiation in SH-SY5Y until stage I (see figure 3.1, step 3) was all-*trans* retinoic acid. In that stage, neurite outgrowth can already be observed. The SH-SY5Y cell line is a heterogeneous culture, since it exhibits different types of cells, the substrate-adherent S-type, the neuroblastic N-type (shown in figure 3.1, step 2) and the morphologic intermediate I-type (figure 3.1, step 1). Only the N-type cells are sensitized for neurotrophic factors, whereas the S-type cells do not respond upon the application of e.g. BDNF (brain derived neurotrophic factor).

Upon the treatment with retinoic acid derivatives, the expression level of several genes was changed (both upregulated and downregulated), amongst them genes involved in cell cycle, differentiation and growth (Hori et al., 2005). In particular, in SH-SY5Y cells, TrkB (tyrosine receptor kinase B) but not TrkA (tyrosine receptor kinase A) is upregulated upon application of all-*trans* retinoic acid (RA). The subsequent addition of BDNF then leads to the phosphorylation of TrkB. Furthermore, the PI3K/AKT signaling pathway is activated by RA due to the promotion of PI3K (phosphatidylinositol-3-kinase), leading to the activation of MAPKs (mitogen activated protein kinases) and finally to cell survival and neurite outgrowth (Pan et al., 2005). As shown in this thesis, the activity of the mitogen activated kinase p38 MAPK was successively increased upon incubation with all-*trans* retinoic acid proportional to the incubation time (see figure 3.5 **A**).

As was shown here, a further increase of p38 MAPK activity was observed upon application of BDNF to the RA-sensitized cells (see figure 3.5 **B**). This could be an indicator that p38 MAPK plays a pivotal role in connecting both differentiation stages I (RA) and II (BDNF). Although the activation mechanisms of the signaling pathways stimulated by RA and BDNF have been extensively studied, the link between the two pathways and the downstream expression of target genes still remains unclear (Nishida et al., 2008).

4.1.2 Phosphorylation of Neurofilaments

The phosphorylation of the NF head domains arises shortly after synthesis in the perikaryon. It is mediated by PKA, PKC and PKN (protein kinases A, C and N) and prevents the subunits from assembly and leading to their disassembly, respectively (Sihag et al., 1999; Mukai et al., 1996; Hisanaga et al., 1990a). The pathologic hyperphosphorylation is caused by the abundant phosphorylation of the C-terminal tail domains, mostly at KSP (lysine, serine, proline) repeats, but also at threonine. The normal non pathologic phosphorylation of the tail domains occurs later upon entry into the axon. Its function is still not elucidated in detail, but it is probably involved in bundle formation and cross-bridge formation between the NF subunits.

However, these two phosphorylation events (on the head and the tail domain, respectively) are putatively interwoven and their acting kinases are orchestrated as was found by Zheng *et al.*, concerning head and tail domain phosphorylation of the medium subunit NFM (Zheng et al., 2003).

4.1.3 Phosphatases in Neurodegeneration

As mentioned earlier in subsection 1.6.4, hyperphosphorylation events are an accompanying effect during neurodegeneration. These abnormally occurring phosphorylation events could be caused by a misregulation between kinases and phosphatases. In the hyperphosphorylation of tau in Alzheimer's Disease (AD), glycogen synthase kinase 3β (GSK) is most likely involved (Hooper et al., 2008). The phosphatase found to be most effective in dephosphorylation of hyperphosphorylated tau was protein phosphatase 2A (PP2A). Its activity was significantly decreased in AD brain (Gong et al., 1993, 1995). The activity of PP2A is regulated by phosphorylation (inactivation) and methylation (activation) and also by two inhibitory proteins termed I_1^{PP2A} and I_2^{PP2A} (Guo and Damuni, 1993; Favre et al., 1994; Li et al., 1995)

Calyculin A (CalA) inhibits the activity of protein phosphatases PP1 and PP2A (Ishihara et al., 1989; Resjo et al., 1999). The application of 10 nM CalA leads to increased phosphorylation of neurofilaments in SH-SY5Y, as was shown by Li *et al.* (Li et al.,

Discussion and Outlook

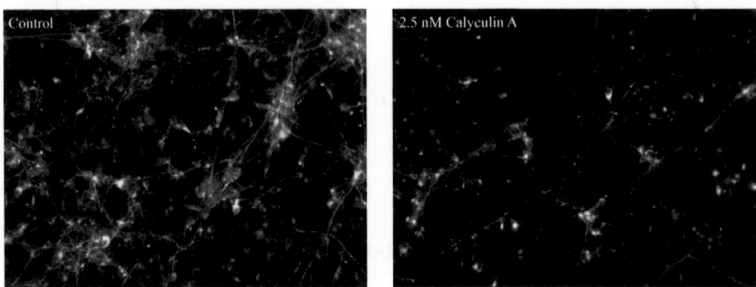

Figure 4.1: The effect of Calyculin A on the O-GlcNAcylated epitope in SH-SY5Y. Cells were treated with either 2.5 nM Calyculin A (CalA) in DMSO (right image) or only DMSO (left image) for 24 h. The O-GlcNAcylated epitope of NFM was immunolabeled with NL6.

2004). This finding was an encouragement to examine the effect of CalA on neurofilament O-GlcNAcylation. Redifferentiated SH-SY5Y were treated with CalA (2.5 nM). Incubation with 10 nM CalA according to Li et al. resulted in an almost complete cell loss. However, in contrast to Li et al., redifferentiated cells were used here.

Nevertheless, this treatment resulted in a decreased signal of NL6 staining, which was most prominent in the axons (see figure 4.1). This reduction in O-GlcNAcylation can be explained by assuming that both PTMs, phosphorylation and O-GlcNAcylation, occur on the same sites and are regulated by their respective transferring enzymes (kinases and OGT) and also by the enzymes mediating their removal (phosphatases and OGase). Indeed, it was shown by co-purifying experiments that OGT (O-GlcNAc transferase) seems to exist in an active complex with PP1, thus acting as a holoenzyme (Wells et al., 2004). This could indicate a preceding removal of a phosphate group by PP1 which is then directly followed by the addition of O-GlcNAc to protect this site from phosphorylation. The inhibition of PP1 (and PP2A) by CalA hampers the removal of phosphate groups, which finally results in this observed decreased in O-GlcNAcylation (see figure 4.1).

Further experiments are performed to get deeper insights into this yin-and-yang-like interplay between phosphorylation and O-GlcNAcylation and their involved respective enzymes.

4.1.4 Responses to Metabolic Stress

Cells are dependent on various metabolites which are used to produce energy, e.g. ATP. Different cels are dependent on different kinds of metabolites. With proceeding specialization (differentiation), cells become highly dependent on a very limited number of metabolites that can be used for energy production (e.g. glucose or pyruvate). This dependency makes them vulnerable to alterations in the metabolite level. Neurons, as highly specialized cells, show various signs of metabolic stress in age induced neurological disorders as is reviewed by Baquer et al. (Baquer et al., 2009).

O-GlcNAc Transferase and AMP Activated Protein Kinase

Glucose deprivation canbe used to induce metabolic stress. Dependent on the duration of this stress, different effects can be observed. Cheung et al. showed that upon short term glucose deprivation for 1-3 h O-GlcNAc Transferase (OGT) activity is decreased, whereas OGT activity is increased after 6-12 h in Neuro 2a cells (Cheung and Hart, 2008). This activation of OGT was suggested to be AMPK (adenosine monophosphate activated protein kinase)-dependent and possibly also mediated by p38 MAPK, either in a direct or an indirect way. Cheung et al. furthermore observed an increase in NFH O-GlcNAcylation following 6-9 h of glucose deprivation. The results shown in figure 4.2 A, exhibit a decreased O-GlcNAcylation of NFM after 15-24 h of deprivation. However, two different neurofilament subunits, namely NFH (Cheung et al.) and NFM (this work) were investigated in two different model systems, namely Neuro 2a, a mouse neuroblastoma (Cheung et al.) and SH-SY5Y, a human neuroblastoma cell line (this work), respectively. Furthermore, Neuro 2a cells were grown in the presence of serum also during the deprivation what resulted in an increase in p38 MAPK activity. In contrast to the findings in this thesis, in which p38 MAPK activity was not affected in SH-SY5Y cells grown in the presence of serum upon glucose deprivation for various durations (9 h and 24 h), as shown in figure 3.5.

AMPK (Adenosine monophosphate activated protein kinase) is known to be activated upon metabolic stress conditions, since it protects the cell from ATP depletion by phosphorylating and thus deactivating enzymes of ATP consuming pathways (Hardie and Hawley, 2001). Additionally, AMPK stimulates catabolic processes by activating glucose uptake (Abbud et al., 2000), glycolysis (Kurth-Kraczek et al., 1999; Marsin et al.,

Discussion and Outlook

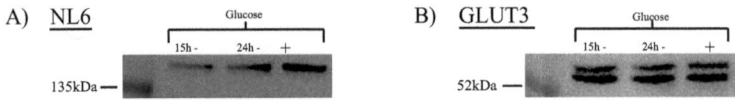

Figure 4.2: Western analysis of O-GlcNAcylation and GLUT3 expression upon glucose deprivation. A The O-GlcNAcylated epitopes of NFM are shown of cells grown in the presence of glucose, and of glucose deprived cells for 15 and 24 h, respectively. In **B** the glucose transporter GLUT3 is shown of cells treated in the same way as described in **A**.

2000), and fatty acid oxidation (Winder and Hardie, 1996) and downregulates anabolic processes, such as fatty acid, cholesterol and protein synthesis, in an attempt to restore cellular ATP levels. Although AMPK has been shown to be expressed in the brain, its effects in neurons is still unknown (Spasic et al., 2009). However, it remains controversial whether AMPK activation following e.g. ischemia, hypoxia or glucose deprivation is beneficial or detrimental (Culmsee et al., 2001; Gadalla et al., 2004; McCullough et al., 2005; Dagon et al., 2005).

p38 Mitogen Activated Protein Kinase and c-Jun N-terminal Kinase

The mitogen activated protein kinase p38 MAPK is involved in the phosphorylation of neuronal proteins upon ischemia and hypoxic conditions as it was shown by Bu *et al.* (Bu et al., 2007). Amongst other proteins, the glucose transporters GLUT1/3 were also found to be phosphorylated by p38 MAPK (Fladeby et al., 2003) upon metabolic stress.

Here, the expression of the glucose transporter GLUT3 was investigated in glucose deprived and control SH-SY5Y cells at differentiation stage II (see figure 4.2 **B**). These cells show a strong expression of GLUT3, that is known to be expressed in neurons. This expression is unaltered upon long term glucose deprivation. Therefore, as a next step, the translocation of GLUT3 to the plasma membrane, upon short and long term glucose deprivation, will be investigated using immunofluorescence.

Based on the findings of Cheung *et al.*, Fladeby *et al.*, Bu *et al.*, Abbud *et al.* and on the results of this work, one can hypothesize the following scene: upon short term glucose deprivation (3-9 h), kinases are activated that stimulate glucose uptake, amongst them AMPK and p38 MAPK. Furthermore, the activity of OGT is increased leading to a protection of neurofilament subunits from hyperphosphorylation. However, upon long term glucose deprivation (>12 h), this endogenous rescue program cannot prevent the cell from hyperphosphorylation events, and finally leads to neurodegeneration.

Future experiments, in which the activity of AMPK in SH-SY5Y cells is investigated, could further enlighten the cellular response to glucose deprivation. The effects of glucose refeeding after the deprivation of glucose on AMPK and also the localization of the glucose transporter GLUT3 will be examined.

Regarding the contribution of other stress activated kinases, phosphorylation of neuron specific c-Jun N-terminal Kinase (JNK) was found to be increased in hypoxic brain (Zhang et al., 2007). JNK is known to be upregulated upon stress, such as upon irradiation with UV light (Adler et al., 1995). To investigate the role of this stress activated kinase in glucose deprived cells, an activity assay was performed as shown in figure 3.6.

A mild increase of JNK activity was observed in both assays upon glucose deprivation. However, one cannot conclude from these assays, whether and to what extent JNK contributes to NF hyperphosphorylation. But because NF subunits are amongst the substrates of JNK, one can assume that JNK contributes to this pathological phenomenon.

4.1.5 Viability of Deprived Cells

The viability of glucose deprived and control cells was checked by staining the mitochondria with the potential sensitive dye JC-1 (see also subsection 5.5.1). In polarized mitochondria the green fluorescent monomers ($\sim$525 nm) form J-aggregates which leads to red-shifting of the fluorescence ($\sim$590 nm). Upon depolarization with oligomycin or similar reagents, the J-aggregates disassemble into monomers. In 15 h glucose deprived cells, the formation of J-aggregates (red) is decreased while the green fluorescence is slightly increased. No J-aggregates are observed in the oligomycin treated cells (see figure 4.3). This indicates that upon deprivation with glucose for 15 h the energy metabolism is affected.

4.1.6 Double Staining of Phosphorylated and O-GlcNAcylated Epitopes

STED microscopy, achieving high resolution for two different dyes has already been implemented using indirect immunostaining (Meyer et al., 2008). The simultaneous observation of the phosphorylated and the O-GlcNAcylated epitopes in the same cell, how-

Discussion and Outlook

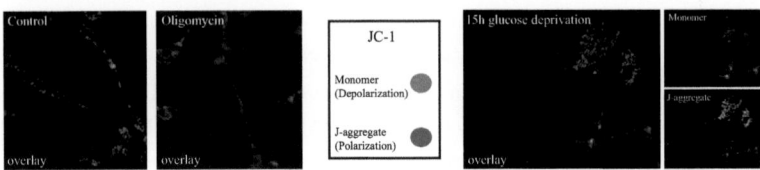

Figure 4.3: Mitochondrial potential in control, oligomycin treated and glucose deprived cells. Cells were incubated with the potential sensitive dye JC-1 for 30 min. Both channels are imaged and overlayed, showing the monomers in green and the J-aggregates in red. In both cases J-aggregates indicate polarized mitochondria and thus viable cells. As a negative control, mitochondria were depolarized with the ATPase inhibitor oligomycin.

ever, requires direct immunofluorescence since all monoclonal antibodies used (SMI34, SMI36 and NL6) were raised in the same host species (mouse). This is a challenge since the amplification effect due to the binding of several secondary antibodies to one primary antibody is lost. To minimize steric interference of the antibodies, the phosphorylated epitopes of NFH and the glycosylated epitopes of NFM were chosen as targets. Therefore, the primary IgG molecules SMI36 and NL6 were directly conjugated to two different dyes as described in subsection 5.5.2. First results are promising, using two spectrally well-separated dyes (OG, Oregon Green 488 and KK114) to exclude unwanted effects such as (F)RET (Fluorescence Resonance Energy Transfer) between the two dyes. As can be seen in figure 4.4, the single stainings (incubation with either SMI36-KK114 or NL6-OG) resemble the double stainings (incubation with SMI36-KK114 plus NL6-OG) what leads to the conclusion that between the two primary antibodies steric hindrance seems to be minimized, as far as can be observed by confocal microscopy.

The application of direct immunofluorescence in high resolution microscopy demands highly photostable fluorophores or sensitive detectors, since the fluorophore density is decreased compared to the indirect method. Subsection 4.3.1 discusses first results of direct immunofluorescence in STED.

4.1.7 Investigation of the Post-Translational Modifications using STED Microscopy

The images of the PTMs were recorded using a supercontinuum STED setup (Wildanger et al., 2008). The resolution achieved was $\sim$30 nm but the recording time for a 10 x 10

4.1.7 Investigation of the Post-Translational Modifications using STED Microscopy

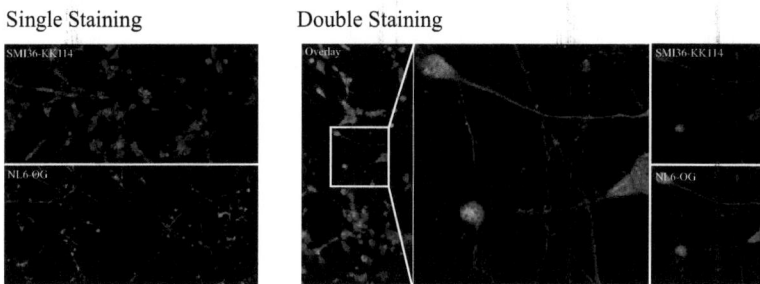

Figure 4.4: Doublestaining of phosphorylated NFH and O-GlcNAcylated NFM. On the left side confocal recordings of single stainings using either SMI36-KK114 (red) or NL6-OG (green) are shown. The simultaneous incubation of both labeled primary antibodies is depicted on the right side, showing both the overlay and the separate channels.

μm^2 image was between approximately 30 min and 45 min. Another STED setup that was built by Dr. Gael Moneron (Department of NanoBiophotonics, MPI for biophysical chemistry) features a slightly reduced resolution enhancement but a highly improved scanning speed (few seconds), due to the implementation of a resonant scanner. To obtain statistically relevant data, future measurements are to be performed with the faster setup.

This setup will help getting deeper insights in the effects of p38 MAPK and JNK inhibitors on the modifications of neurofilaments under glucose deprivation and normal culture conditions.

Discussion and Outlook

4.2 Incorporation of Labeled Lipids in the Plasma Membrane

Two techniques were applied in this thesis to introduce labeled lipid analogs into the plasma membrane. The first method is based on the generation of liposomes which then fuse with the plasma membrane. The second technique, which was improved for this project in this thesis, makes use of the complex formation of BSA and lipids in aqueous solutions.

4.2.1 Fusion of Liposomes with the Plasma Membrane

The use of liposomes allows one to introduce lipids in cell membranes. Their composition has to resemble the proportion of lipids present in the plasma membrane, or else the membrane properties are changed. However, the amount of the finally incorporated lipids could only be controlled to some extent since the number of fusion events strongly varied between the cells. Moreover, the size of the liposomes which determined the fusion capability was distributed statistically. That resulted in a proportion of large liposomes which fuse to a much lower extent but contributed to high background by adhering to the glass surface of the coverslip.

The presence of calcium cations in the fusion buffer was an absolute prerequisite for liposome fusion. Different calcium concentrations were tested ranging from 1 mM up to 100 mM $CaCl_2$. While only few liposome fusion events took place when using buffers ranging from 1 mM to 10 mM, multiple events could be observed with concentrations up to 100 mM. Side effects of the high Ca^{2+} ion concentration on the properties and function of the membranes can occur, such as aggregation of vesicles containing a net negative charge. Furthermore, the Ca^{2+}-dependent protease calpain cleaves, amongst other targets, spectrin, a membrane skeletal protein. Its cleavage leads to the disruption of the anchoring of cytoskeletal elements, in particular of actin, to integral membrane proteins and to the formation of so called blebs, usually finally leading to the death of the cell (Castillo and Babson, 1998). This loss in membrane integrity can be amplified by the activation of phospholipases which leads to the hydrolysis of phospholipids (Farber, 1990; Nicotera et al., 1990).

However, since the fusion of the liposomes with the plasma membrane was carried out at 4 °C, the activation of these enzymes during this incubation step was reduced to a minimum. After fusion of the liposomes with the plasma membrane, the Ca^{2+} buffer was replaced to minimize harmful effects, in particular on the sheets, which were generated directly thereafter.

The incorporation of liposomes changes the overall lipid content of the plasma membrane, since an extra amount of lipids is introduced. Also for this reason it was crucial to adjust the liposome composition to the one of the plasma membrane. Still the amount of fusion events per cell remained uncontrollable, meaning that the labeling density greatly varied between the cells.

4.2.2 Plasma Membrane Sheet Generation

The application of a brief ultrasonic pulse disrupted the upper plasma membrane, without altering the lower membrane attached to the glass surface. These so called membrane sheets were not homogeneously distributed, but were only found in an actinism-like pattern in the proximity of the main impact of the ultrasonic pulse. In the center of the main impact however, hardly any cellular component was left, whereas in the further surrounding the cells were completely intact, or still possessing upper membrane leftovers, respectively. This significantly restricted the number of intact sheets to a small zone.

Since the concentration of Ca^{2+} ions had to be kept low for the plasma membrane sheets, the cells for the sheet generation were transferred into an EGTA-containing potassium-glutamate buffer that mimicked the cytosol. This buffer was also used for the dynamic measurements with the STED-FCS setup. But as already mentioned in subsection 3.2.2, the loss of integrity of the sheets significantly constricted the time window for the dynamic recordings, so that measurements had to be completed within $\sim$ 10 min including the mounting of the sample and finding the membrane sheets. Therefore, to gather statistically relevant data, this time window was simply too small.

4.2.3 Serum Albumins - Lipid Carriers in the Blood

Serum albumins are the most abundant plasma proteins in all mammals and are essential for maintaining the osmotic pressure needed for the proper distribution of body fluids between the intravascular compartments and tissues. Approximately 75% of the total

Discussion and Outlook

colloid osmotic pressure is related to albumin. Furthermore, albumin plays an important role in maintaining homeostasis within the body (Hankins, 2006).

The technique applied here to introduce the labeled lipids made use of the ability that serum albumin can unspecifically bind fatty acids and other hydrophobic molecules, such as steroid hormones and hemin. As can be seen in figure 3.9, BSA (bovine serum albumin) carries seven arachidonic acid molecules in hydrophobic pockets (Petitpas et al., 2001). By that the fatty acids and other hydrophobic molecules can be transported in aqueous solutions.

Upon complex formation between BSA and monoacyl-phospholipids in aqueous solutions, the fatty acid resides in one of the hydrophobic pockets of the BSA molecule whereas the hydrophilic head domain directly interacts with the environment. Thus, micelle formation is eliminated. 100 nmol of both the lipid and the BSA molecules were used for complex formation. However, the two fatty acids of diacyl-phospholipids are most unlikely complexed by one single BSA molecule, since the hydrophobic pockets of BSA are in a sterically unfavorable position in respect to each other to allow coinstantaneous protection. Leaving one of the two acyl chains unprotected hampers the incorporation of the phospholipid molecule into the plasma membrane. One could assume that this second fatty acid residue could be protected by another BSA molecule. Since incorporation was still observed but to a lower extent than compared to the monoacyl-phospholipids, the conclusion was drawn that the amount of BSA molecules does not depend on the number of phospholipid molecules, but on the amount of their acyl chains. As a result, 100 nmol diacyl-phospholipids were incubated with 200 nmol BSA molecules, resulting in an incorporation efficiency comparable with monoacyl-phospholipids.

Using the higher BSA amount (200 nmol) also for monoacyl-phospholipids did not alter their incorporation efficiency. However, both monoacyl-and diacyl-lipids were complexed with BSA using a 1:1 (number of acyl chain:number BSA molecules) ratio.

4.2.4 Proper Incorporation of the Labeled Lipids in the Plasma Membrane

The membrane affinity of the labeled lipid analogs was checked by back exchange experiments with BSA washing. Lipids, which were not properly incorporated into the plasma membrane, could be removed completely by BSA washing. Figure 4.5 depicts, how a

4.2.4 Proper Incorporation of the Labeled Lipids in the Plasma Membrane

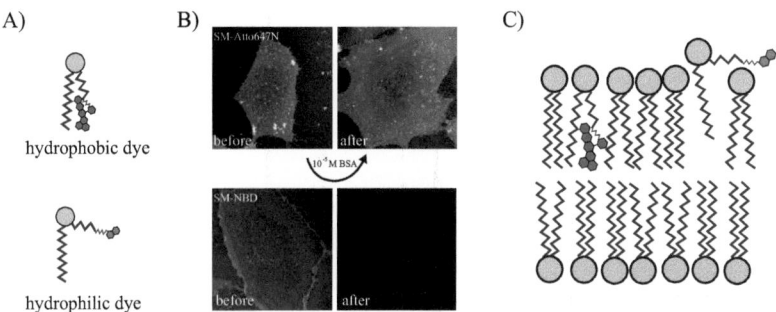

Figure 4.5: **The influence of the hydrophilicity of the dye on the lipid incorporation into the membrane.** Two labeled analogs of SM are shown in **A** with one hydrophilic and one hydrophobic dye, respectively. Upon incorporation into the plasma membrane BSA back exchange experiments are performed, as shown in **B**. In **C** a model of the membrane affinity for hydrophobic and hydrophilic lipids shows both lipids from **A** incorporated into a lipid bilayer in an aqueous surrounding.

label could affect the membrane anchoring (here: a hydrophilic and a hydrophobic dye as acyl chain replacement). The sketch of the NBD-labeled SM analog in figure 4.5 **A** was drawn in accordance with van Meer *et al.* (van Meer and Liskamp, 2005). As depicted in figure 4.5 **B**, the hydrophobic dye (here: Atto647N) when used as an acyl chain replacement showed no significant decrease in membrane staining after the BSA washing, whereas the hydrophilic dye (here: NBD) could almost completely be depleted from the plasma membrane. The sketch in **C** shows a model in which the two labeled lipids are incorporated in a lipid bilayer surrounded by an aqueous solution. The hydrophilic dye (green) situates itself at the water-lipid interface, and can therefore be removed a lot easier from the membrane by BSA than the hydrophobic dye (red) which is embedded in the membrane interior, possibly also establishing membrane anchoring.

Dyes attached to the hydrophilic head domains of the lipids had less influence on the membrane anchoring, which is in fact primarily provided by the fatty acid tail domains. Nevertheless, the diffusion of differently labeled lipid analogs was investigated, differing in the label's polarity, its labeling position and in the amount of acyl chains. The hydrophobic dye Atto647N was performing best as a label, independent from its labeling position (head or tail), and was therefore chosen for all FCS experiments. Moreover, Atto647N is remarkably photostable and no toxicity was observed, neither from the dye itself nor from radical formation during data recording.

4.2.5 Diffusion of Phospholipids with Saturated and Unsaturated Fatty Acids

Most synthesized labeled lipid analogs contained saturated fatty acids. As it is known from for example crystallographic studies, phospholipids with only saturated fatty acids are densely packed in the membrane whereas unsaturated (mono- or poly-) fatty acids feature one or more double bonds (Small, 1984). In *trans*-configuration, their shape is comparable to the straight saturated fatty acids whereas a *cis*-configuration causes the chain to bend. One *cis*-double bond results in a kink while two or more result in a bend or a hook. Introducing a kink in the fatty acid chain of a phospholipid limits its ability of being densely packed. Most unsaturated fatty acids are found in the *cis*-configuration.

Phospholipids are in general susceptible to hydrolysis. Moreover, phospholipids containing unsaturated fatty acids can undergo oxidative reactions, meaning that preparative procedures and also storage conditions have to be carried out in an oxygen-deficient atmosphere. Here, all lipids used for liposome preparation were saturated, since during sonication and homogenization oxidation can take place to a high extent.

To investigate the diffusion behavior in the plasma membrane of unsaturated phospholipids, the mono-unsaturated Atto647N-labeled analog of PE, DOPE, was introduced via BSA complex formation. During storage and handling, direct exposure to air oxygen was avoided by working under a constant flow of nitrogen. However, the incubation of the BSA-lipid complex with the cells was carried out under normal conditions, it was assumed that BSA acted protectively against oxidation by enclosing the acyl chains within its hydrophobic pockets. Nevertheless, it was found that DOPE behaved similar to its saturated counterpart PE. Its diffusion could be described by a single species fit (similar to PE). However, for simplicity in preparation and handling, saturated phospholipid analogs were used for most of the experiments.

4.2.6 Outer or Inner Leaflet?

The lipid bilayer of the plasma membrane consists of two monolayers, the outer and the inner leaflets which face the extra- and the intracellular sides respectively. The asymmetric lipid composition between the two leaflets is functionally highly important. Upon incorporation of the labeled lipid analogs, the question arose where within the plasma membrane they are localized. The thickness of the lipid bilayer is about 5 nm. Therefore, the two leaflets cannot be discriminated with the axial resolution of the STED-FCS setup. Nevertheless, the diffusion behavior of the introduced lipids could vary depending on their localization. The trapping observed for sphingolipids was found to be cholesterol mediated, which is found in both leaflets.

As depicted in figure 3.9 **A**, upon the generation of liposomes, the labeled lipids are distributed in both liposomal leaflets. Fusion of these liposomes with living cells also leads to the distribution of labeled lipids in the outer and inner leaflet of the plasma membrane. On the contrary, the introduction of BSA-complexed labeled lipids only results in an incorporation in the outer leaflet of the plasma membrane. Lipid exchange between the two leaflets, also known as flip-flop, occurs very rarely (Devaux, 1993), unlike the horizontal exchange within one leaflet.

It is therefore important to keep in mind that all acquired data of the labeled lipids, which were incorporated via BSA-complexing, originate from the outer leaflet of the plasma membrane.

4.2.7 Cholesterol Depletion

The distribution of sterols among the membranes of eucaryotic cells is heterogeneous. The plasma membrane, for instance, contains a high concentration of sterols, whereas the intracellular membranes of mitochondria, the endoplasmic reticulum and other organelles have little sterols. Here in this work, two different methods were used to deplete the endogenous cholesterol from the plasma membrane: enzymatically by cholesterol oxidase (COase) and by host-guest complex formation by β-cyclodextrin (β-CD).

The oxidoreductase COase converts cholesterol to cholest-4-en-3-one in a two step reaction (Liu et al., 1996). A byproduct of this reaction is the generation of H_2O_2 which, as a potent oxidant, could additionally lead to stress. However, the viability of the cells was judged by their morphology and constantly checked. Furthermore, the efficiency of COase in PtK2 cells could not be enhanced, since an increase in concentration and/or incubation time led to cell death.

The oligosaccharide β-CD can act as a cholesterol acceptor or donor according to the

Discussion and Outlook

ratio between the host and the guest molecules. Furthermore, the concentration, incubation time and cellular responses greatly vary between the different cell types and tissues (Zidovetzki and Levitan, 2007). Therefore, different concentrations and incubation times were tested, and their depletion efficiency judged by phallotoxin staining of the F-actin cytoskeleton. The influence of the endogenous cholesterol level on the strengthening of the stress fibers was used as an indicator for effective cholesterol depletion. The incubation with 10 mM β-CD was most efficient.

As can be seen in figure 3.16, the treatments with COase and β-CD varied concerning their effect on the actin cytoskeleton. This could be an indication for a more efficient cholesterol depletion by β-CD. This different depletion efficiency was also reflected in the obtained FCS data. The anomalous diffusion behavior of sphingolipids could be almost completely abolished upon the treatment with β-CD, whereas the diffusion upon cholesterol depletion with COase partly remained anomalous (compare with subsection 3.2.5).

Furthermore, in addition to its role in lipid microdomains, Orr *et al.* suggested that cholesterol plays a potent modulatory role in the mobility of transmembrane receptors (EGFR) in the membrane (Orr et al., 2005). It could thereby provide a possible mechanism by which cholesterol modulation affects receptor activation.

4.2.8 The Interactions of the Cytoskeleton with Constituents of the Plasma Membrane

The cytoskeletal elements establishing and preserving the cellular shape and allowing for its motility need to be connected to the plasma membrane. A skeletal scaffold protein, spectrin, underlies the intracellular side of the plasma membrane. It acts as an anchoring point for other scaffolding proteins, such as the actin cytoskeleton via linking proteins such as ankyrin and protein 4.1. In certain types of brain injury such as diffuse axonal injury, spectrin is irreversibly cleaved by the proteolytic enzyme calpain, destroying the cytoskeleton (Buki et al., 2000). Spectrin cleavage causes the membrane to form blebs and ultimately to be degraded leading to the death of the cell (Castillo and Babson, 1998). As mentioned before, this bleb formation can also be introduced by high Ca^{2+} concentrations which activate calpain. Further, it was shown that the interaction of phospholipid monolayers with erythrocyte and brain spectrin was cholesterol dependent (Diakowski et al., 2006).

Phosphatidylinositol-(4,5)-bisphosphate (PIP$_2$), which is thought to accumulate in 'lipid rafts', is the major actin cytoskeleton regulator (Caroni and Golub, 2002) which has been shown to be delocalized from the membrane with cholesterol depletion (Pike and Miller, 1998). The sequestration of PIP$_2$, e.g. by cholesterol depletion, alters the organization of actin and inhibits the lateral diffusion of membrane proteins (Kwik et al., 2003). The motion of membrane proteins is confined by the cortical, membrane-associated F-actin (Yechiel and Edidin, 1987; Edidin and Stroynowski, 1991). The actin strands in the cytoskeleton network are thought to sterically interact with the cytoplasmic tails of these transmembrane proteins, thus confining them into microdomains (Orr et al., 2005; Edidin, 2001; Saxton and Jacobson, 1997). The immobilized membrane-associated proteins that interact with the cytoskeleton could form fences and pickets within the plasma membrane (Fujiwara et al., 2002; Tang and Edidin, 2003).

Further STED-FCS experiments are focused on the influence of spectrin and actin on the lipid diffusion, in particular of the behavior of sphingolipids. This includes experiments in which the actin cytoskeleton is either stabilized (phalloidin) or destabilized (latrunculin A). First results indicate a loss of trapping upon actin destabilization. The modification of the spectrin cytoskeleton involves the inhibition or activation of calpain or overexpression of spectrin by transfection of plasmid DNA.

4.2.9 Temperature Dependence on Lipid Diffusion

The temperature was a not negligible parameter, since it had to be balanced out between the needs of the cells and the optical performance of the setup. Therefore, all measurements were performed at either 37 °C or 27 °C. Besides the faster lipid internalization and faster cholesterol recovery after depletion (see section 3.2.5) at 37 °C compared to 27 °C leading to a smaller acquisition time window, the lipid dynamics were hardly altered. At 27 °C an overall slight slowdown of the free diffusion could be observed (diffusion coefficient $D_{27°C} \approx 5 \times 10^{-9}$ cm^2/s compared to $D_{37°C} \approx 6 \times 10^{-9}$ cm^2/s) whereas the sphingolipid trapping remained unchanged (e.g. SM-Atto647N $\tau_{trap} \approx 10$ ms and $A_2 \sim 60\%$ for both temperatures 27 °C and 37 °C).

However, the optical performance of the STED-FCS setup in terms of optical aberrations was slightly better at lower temperatures. Therefore and for experiments requiring

Discussion and Outlook

statistical evaluation, most data were acquired at 27 °C. That was also true for the back exchange experiments and the FRAP measurements performed with a confocal laser scanning microscope.

4.2.10 Potential Artifacts due to the High Intensities of the STED and Excitation Lasers

Since the investigation of the lipid dynamics was performed on living cells, the possibility exists to induced adverse effects with the excitation and the STED lasers. These effects could be photobleaching of the fluorophore (Eggeling et al., 1998), radical generation (Girotti, 2001; Ayuyan and Cohen, 2006) and local heating by light absorption of the surrounding water, lipids and proteins (Schönle and Hell, 1998).

The STED laser wavelength used for these experiments is in the near-infrared (770-780 nm) and is compatible with living cells. The pulsed STED intensity of up to 6 GW/cm^2 (corresponds to an incident laser power of > 100 mW) is by a factor of $\sim$ 100 lower (since the applied STED pulses are 100-1000 times shorter) compared to the intensities used in multi-photon microscopy, which is a routinely used technique for live cell observations (Denk et al., 1990). Taking this into account results in a comparable pulse energy, or at the most by a factor of ten larger in the STED case. However, all measurements were completed before morphological changes occurred which was constantly checked by the cell's appearance in transmission light microscopy.

The local heating of the cells in the sample can be ruled out, since the heat dissipation in aqueous solutions surrounding the cells has been shown to be effective, even at these applied laser powers (Schönle and Hell, 1998). Moreover, the STED beam is, in contrast to the focused beam in multi-photon excitation, spread out into a doughnut which is $\sim$4-5 times larger and therefore guarantees a more effective heat dissipation.

The generation of radicals from excited fluorophores cannot be excluded, but is minimal due to the low concentration of introduced labeled lipids. Light induced radical formation can lead to lipid peroxidation, in particular of unsaturated lipids, thus leading to a change of fluidity (Girotti, 2001). Furthermore, cells are provided with natural radical scavenging systems which can account for low concentrations of radicals (Ayuyan and Cohen, 2006). Since no change of fluidity was observed in the experiments, it was concluded that the amount of radicals possibly generated was tolerable for the cells.

4.2.10 Potential Artifacts due to the High Intensities of the STED and Excitation Lasers

The photobleaching induced by the STED beam could be excluded, since the de-excitation rates are more efficient than the competing photobleaching rates in the spot periphery (Hell, 2007; Eggeling et al., 1998). However, photobleaching by the excitation light, in particular for the confocal FCS experiments, plays a non negligible role. This can be explained by the prolonged transition time, since the confocal detection volume is much larger than the effective detection volume in the STED mode. That means that the probability of photobleaching defines the observation time of a molecule. Furthermore, to detect trapping events, the observation time has to be significantly longer than the molecule's trapping time. By using the photostable dye Atto647N, photobleaching by the excitation beam could also be extensively excluded under the circumstances used for the measurements.

Discussion and Outlook

4.3 Specific Labeling in High Resolution Microscopy

Immunofluorescence is a wide spread method used throughout the field of life sciences. As the size of the IgG molecule (∼10 nm) was by an order of magnitude away from the diffraction barrier, its size was not regarded as a problem. However, with STED and other high resolution microscopy techniques overcoming this barrier, alternative labeling strategies have to be considered. Fluorescent proteins are smaller (∼3x3x4 nm^3) but they cannot replace affinity markers which can specifically detect, for instance, PTMs. In the following, some adaptations of the label to high resolution microscopy are discussed considering a decrease of the detection complex and an increase in photostability.

As already mentioned in the introduction, the phostostability of the fluorescent dyes is a pre-requisite since the resolution obtained with STED is theoretically infinite and depends on the STED laser's intensity. For the here so far introduced techniques (STED-FCS and STED in imaging), the Atto dyes as well as some common dyes, such as Oregon Green 488 and Alexa 488 can be used.

4.3.1 Alternative Affinity Markers

Imagine a filament (e.g. a microtubule filament) with a diameter of ∼25 nm detected by specific IgG molecules distributed around the tube. These IgG molecules increase the outer diameter of the filament by 10 nm each, resulting in a final outer diameter of 45 nm. The usage of indirect immunofluorescence would lead to 65 nm. All these calculations are theoretical and are based on the structural informations obtained from Murphy *et al.* (Murphy et al., 1988). The effective size of the antibody-antigen complex including their hydration could differ from these values. However, what can be concluded from these assumptions is that the detecting complex lies in the same order of magnitude as the structure itself.

The direct immunofluorescence allows a reduction of the antibody-target complex (see figure 1.2) by a factor of ∼2. First examples, in which a primary IgG molecule was directly conjugated to a fluorophore with an average DOL (degree of labeling) of 3 was successful in generating STED images. Figure 4.6 shows an example of indirect and direct immunofluorescence (here: Dyomics 495). The images acquired of the direct

4.3.1 Alternative Affinity Markers

and indirect staining of vimentin differ in their number of counts, which is decreased in the direct staining (right image) due to the lack of signal amplification by the secondary antibody. This limitation, however, can be overcome by the implementation of more sensitive detectors (avalanche photo diodes, APD).

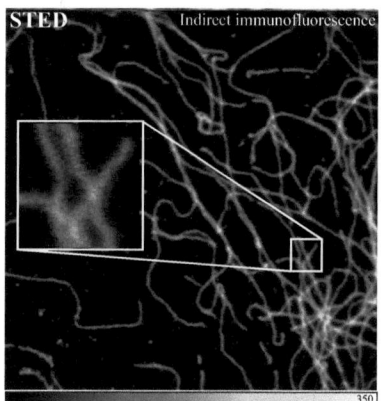 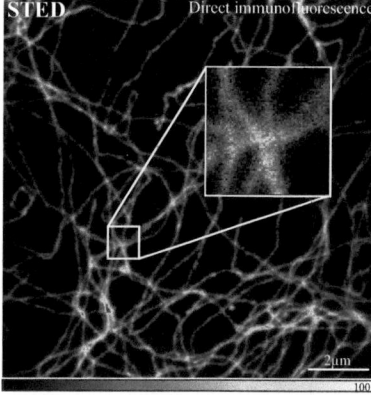

Figure 4.6: Indirect and direct immunofluorescence of vimentin recorded with a STED microscope (built by Dr. Gael Moneron). The left image, showing the indirect immunostaining of vimentin, using a dye-conjugated secondary antibody. On the right side, the directly conjugated primary antibody stains the same structure. The colormaps below the images, which are a measure for the photon counts, depict different maximum values corresponding to the different dye-density in both methods. The direct labeling lacks the signal amplification by the binding of several secondary antibodies to one primary. The edge length of the inset is 1μm.

To further reduce the size of the detection complex, Fab fragments were generated by successive digestion with immobilized papain (Pierce, Rockford, IL, USA) according to the manufacturer's protocol. Briefly, the IgG molecules were dialyzed against 20 mM Na_2HPO_4/10 mM EDTA for 24 h and incubated with the immobilized papain in digestion buffer (20 mM cysteine-HCl, 10 mM EDTA, pH 7.0) at 37 °C for 48 h. The Fab fragments were then separated from undigested whole IgG and Fc by a protein A column. The Fab containing fraction was dialyzed against PBS, pH 7.4, and the size confirmed by SDS PAGE (50 kDa). The Fab fragments were labeled with NHS ester dyes and purified with a G25 Sephadex column according to the protocol described in 5.5.2. But because of the high loss of material due to incomplete digestion and recovery after purification, and with

Discussion and Outlook

a poor DOL of ≤ 1, this method was not practicable.

Another kind of affinity markers are the so called aptamers. These are DNA or RNA molecules which, similar to antibodies, sterically bind and detect their targets (Nitsche et al., 2007; Breaker, 2004). Their size ranges between $\sim$25-60 nt (James, 2001).

The aptamers used here, were produced from the company Aptares (Aptares, Mittenwalde, Germany) using the purified NFH protein from Progen (Heidelberg, Germany) as matrix. The conjugation with Atto647N was carried out by binding of the NHS ester to an introduced amino group. The excess free dye was removed by dialysis directly prior to the experiments. Since aptamers are rather small (James, 2001), they could be interesting tools for live cell investigations.

The usage of aptamers in live cell experiments required a method capable of introducing the molecules through the plasma membrane into the cytosol without damaging the cell. In principle microinjection is a versatile method, but, however, not practicable since every cell had to be injected manually. Moreover, the cells had to be incubated at 37 °C and 5% CO_2 for up to several hours to allow aptamer binding to the NFH proteins. Using Nanofectin (PAA, Pasching, Austria), a reagent normally used for transfection, succeeded in introducing aptamers inlive cells, as shown in first tests (see figure 4.7).

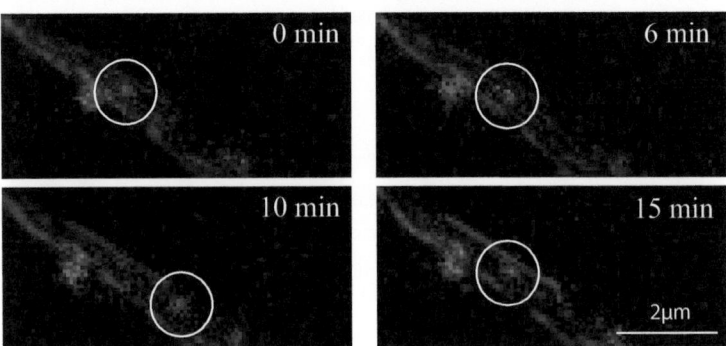

Figure 4.7: NFH labeling using DNA aptamers conjugated to Atto647N. SH-SY5Y cells at differentiation stage II were incubated with Atto647N-labeled aptamers (green). Mitochondria are stained with the mito tracker dye $DiOC_6$ (red). The images were recorded with a standard confocal microscope. 3D stacks were taken every minute for a total duration of 15 min. All images shown are maximum projections of the third axis.

So far the effective use of these markers in high resolution tracking experiments was not successful, since the photostability of the used dye (in both STED and confocal experiments) was not sufficient to allow tracking for long term studies as such as neurofilament subunit motility (several hours). However, an example is shown in figure 4.7 which was recorded at the confocal laser scanning microscope Leica TCS SP5 (see subsection 5.6.2) for 15 min. The aptamer-Atto647N labeled NFH is shown in green, whereas the mitochondria are stained with the mito tracker $DiOC_6$ (shown in red). The cells are mounted on a homebuilt microscope chamber and imaged in prewarmed HDMEM (HEPES buffered DMEM without phenolred, according to the medium used for STED-FCS recordings). A 3D stack is acquired every minute. All images in figure 4.7 are maximum projections of the third axis (axial, z), whereas the fourth axis (t, time) is shown in steps of 1 min. The white circle tracks a moving aptamer-NFH complex in all images.

However, similar to experiments, in which proteins are tagged with fluorescent proteins, one cannot exclude the possibility that the bound aptamer does not hamper or alter the movements and motility of the labeled NFH. Furthermore, as can be seen from these images, the signal-to-noise ratio (SNR) is rather low in the aptamer channel. To improve the SNR, an alternative strategy is discussed in the following subsection.

4.3.2 *in vitro* Labeling of Recombinant Proteins

As discussed in the first section, many life cell applications, such as protein tracking, require a labeling of the structure which does not significantly interfere with the molecule's behavior. For tracking experiments, no dense labeling is necessary in contrast to imaging. However, there is the need to follow a fluorescently tagged molecule over a long time. In that case it also differs from FCS, which records the intensity fluctuations of molecules diffusing in and out of the focal volume for only short time scales. Affinity markers, such as antibodies, often have the tendency of complexing their target molecules when used in non fixed, life cell approaches what could lead to undesirable side effects. The aptamers, on the other hand, feature a rather low SNR and require moreover an additional incubation step to allow binding to the NFH subunits.

Therefore, according to the approach realized for the labeling of lipids, recombinant proteins were labeled *in vitro* and introduced into the cell. Also for this technique, microinjection was not performed here. The neurofilament triplet proteins, like most other

Discussion and Outlook

cytoskeletal proteins, are highly stable. Therefore, NF subunits, whose functionality was already proven by the manufacturer, can be purchased as recombinant or purified proteins (Progen, Heidelberg, Germany). Since the primary structure of the NF subunits is known (shown in the supplementary), the position and number of the free primary amino groups for NHS ester conjugation or free thiol groups for maleimide conjugation could be determined. As a first test, Atto647N-NHS was chosen, since it did not show any toxic effects in live cells (it was also chosen for the lipid experiments).

The investigation of neurofilament subunit transport will be performed using *in vitro* labeled recombinant (or purified) NFH. Since it is still an open question whether NF subunits are transported along the axon as single subunits or oligomers, using the STED setup of Gael Moneron could be a sufficient tool to investigate this question. By modifying presumably involved enzymes, their effects on the mobility of NFH could directly be observed.

As a first example, Atto647N-conjugated NFH (here:green) was introduced in redifferentiated SH-SY5Y cells using Pro-Deliverin (OZ Biosciences, Marseille, France) and DiOC$_6$ stained mitochondria (red), as shown in figure 4.8. This image was taken from a 4D stack (x, y, z, time) recorded with a confocal microscope (Leica TCS SP5) for several minutes.

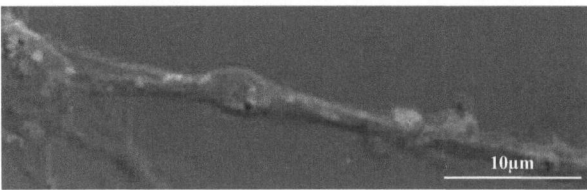

Figure 4.8: *in vitro* **labeling of NFH with Atto647N.** Confocal recording of the *in vitro*-Atto647N-labeled NFH subunits introduced into redifferentiated SH-SY5Y cells. As a viability control, the mitochondria are visualized with DiOC$_6$ (red) and, additionally, a DIC image (gray) was taken.

The proper incorporation into the neurofilament network will be examined by using siRNA (small interference RNA) to downregulate NF subunits and rescuing these cells with the addition of *in vitro* labeled protein. Since NFL is required to establish the neurofilament protofilament, experiments using NFL-siRNA and NFL-Atto647N, respectively will be performed to examine proper functionality of the labeled proteins.

4.3.3 N-V Centers: Extremely Stable Labels in High Resolution Microscopy

Nanodiamonds (NDs) containing N-V (nitrogen vacancy) centers are fluorophores with remarkable properties: high photostability and a large quantum yield. Therefore, they are highly promising for the use in high resolution optical microscopy as was shown recently (Rittweger et al., 2009). To open them to the field of life sciences, biofunctionalization is an absolute pre-requisite. Recent studies have proven that fluorescent and photoluminescent diamonds that were taken up by cells showed no sign of toxicity (Chao et al., 2007; Faklaris et al., 2008). However, in these experiments, the diamonds were not specifically targeted to distinct cellular structures.

The Nitrogen-vacancy center (N-V center) is described as a point defect in the diamond lattice. It consists of a next neighbor pair of a nitrogen atom, which replaces a carbon atom, and a lattice vacancy, forming a fluorescent system which is bright enough for individual N-V center detection. These resonances (fluorescence) in NDs are electron spin related phenomena, such as quantum entanglement, spin-orbit interaction, and Rabi oscillations. The electron spins of individual N-V centers can be manipulated at room temperature by applying light, magnetic, electric or microwave fields (or their combinations). The application of these N-V centers ranges from electronics and computational science including spintronics, quantum cryptography, quantum computing, etc. to optical microscopy.

Two charge states, N-V^0 and N-V^- of this defect are known from spectroscopic studies including absorption (Davies and Hamer, 1976; Mita, 1996), photoluminescence (Iakoubovskii et al., 2000), electron paramagnetic resonance (EPR) (Loubser and Wyk, 1978; Redman et al., 1991) and optically detected magnetic resonance (ODMR) (Gruber et al., 1997). The N-V^0 state is defined as the neutral charge state, whereas the N-V^- state is the negative charge state.

Three of the five valence electrons of the nitrogen atom are covalently bond to the carbon atoms, leaving the remaining two non-bonded. The vacancy has three unpaired electrons with two of them forming a quasi covalent bond and the other one remaining unpaired. However, the three unpaired vacancy electrons continuously exchange their

Discussion and Outlook

roles. The symmetry is trigonal pyramidal (C_{3V}). In the N-V^0 five electrons contribute to this system, whereas for the N-V^- six electrons are needed.

The neutral charge state N-V^0 is paramagnetic. Only recently it was possible to detect them with EPR, since optical excitation was required to bring the N-V^0 defect into the EPR-detectable excited state (Felton et al., 2008). The better-documented state, however, is the negative charge state N-V^-. Its extra electron is located at the vacancy site forming a spin S=1 pair with one of the vacancy electrons. These spin states can be optically manipulated and readout. As in N-V^0, the vacancy electrons continuously exchange roles, thus preserving the overall trigonal symmetry.

To make nanodiamonds usable for the life sciences community, the fluorescent nanodiamonds (FNDs) needed to become biocompatible and functionalized. Here, the N-V^- FNDs, ranging in size between 20 nm and 50 nm, were used for the following experiments.

One possibility to specifically label structures with an FND is the direct attachment of a biotin molecule to the diamond. The first batch of FND-Biotin (made from detonation FNDs) was kindly provided by Dr. Anke Krüger (Nanocarbon Materials Group, Institut für Organische Chemie, Christian-Albrechts-Universität Kiel). U373MG cells (human glioblastoma cell line) were immunolabeled using a biotin conjugated anti-GFAP (glial fibrillaric acidic protein) antibody. After incubation with neutravidin (Invitrogen, Carlsbad, CA, USA), the biotin-avidin complex was fixed with PFA. Subsequently, severe extraction with Triton X-100 had to be performed, accordant to the introduction of antibody-conjugated Quantum Dots (Medda et al., 2006), to introduce the biotin-FND complexes. However, the yield of NDs bearing a N-V center was rather low. Furthermore, they exhibited photobleaching upon exposure to the STED beam (Kyu-Young Han, private communications). But still, in confocal recordings (Leica TCS sp5, see 5.6.2) binding of the FND to a filamentous structure, GFAP, could be observed (see figure 4.9 A).

Furthermore, carboxylated FNDs (mean diameter $\sim$35 nm) from Dr. Huan-Cheng Chang were functionalized with biotin in the group of Dr. Vladimir Belov. Since these FNDs were larger than the ones from Anke Krüger, they were not used for the labeling of intracellular structures, such as GFAP, but for structures on the cell surface. PtK2 cells were transfected with a plasmid encoding a GPI-avidin fusion protein (see subsection 5.3.8). The biotin-FNDs could directly be incubated with the avidin-expressing cells. As

4.3.3 N-V Centers: Extremely Stable Labels in High Resolution Microscopy

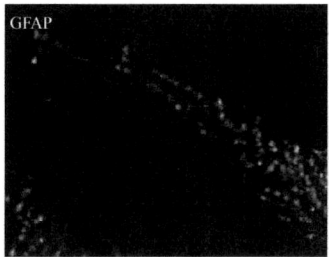
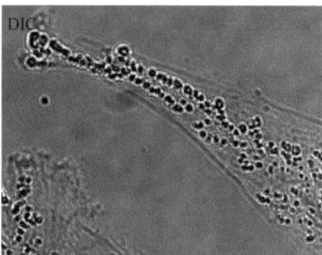

Figure 4.9: Biofunctionalized FNDs. N-V$^-$ FNDs functionalized with biotin from Anke Krüger are incubated with U373MG cells, immunolabeled with anti-GFAP-Biotin mouse IgG. Unlabeled neutravidin sandwiches the biotin-conjugated antibody and the biotin-conjugated FND. The filamentous structure of GFAP is clearly marked by the nanodiamonds, as can be also seen in the DIC image. The images were recorded using standard confocal microscopy.

a positive control, Atto565-biotin (Atto-Tec, Siegen, Germany) was used. However, so far no specific labeling with the 35 nm FNDs could be obtained. The influence of the size of the FNDs on affinity binding is discussed below.

A non negligible fact is the huge size of the FNDs, 10-50 nm, making the conjugation to biomolecules challenging. An average immunoglobulin molecule is $\sim$ 10 nm in diameter. In cooperation with Dr. Vladimir N. Belov's group, different strategies were pursued. The direct conjugation of the FND to the immunoglobulin molecule could lead, in particular for the large particles, to interference with the antibody's affinity to its targets due to rigidization of the antibody and steric reasons (see figure 4.10). The non-covalent binding between antibody and its target is strong for two equally sized binding partners (e.g. primary and organic dye-conjugated secondary antibody), but is weak for partners greatly differing in size (e.g. primary and FND-conjugated secondary antibody).

The rigidity of this complex can be reduced by introducing a linker between the FND and the antibody. Furthermore, due to the increased flexibility, the forces exerted on the binding site are decreased. Several linkers varying in their length were tested. These linkers were attached to the diamonds and via a second, bifunctional cross linker conjugated with the antibodies.

Such a bifunctional cross linker is succinimidyl-4-(*N*-maleimidomethyl)cyclohexane-1-carboxylate (SMCC), which contains an amine-reactive *N*-hydroxysuccinimide (NHS ester) and a sulfhydryl-reactive maleimide group (Pierce Biotechnology, Rockford, IL,

Discussion and Outlook

USA). Since the NHS ester is prone to fast hydrolysis in aqueous solutions, first of all, SMCC was conjugated to the primary amines of the antibodies at pH 7.4 (tolerance range pH 7-9) using a five- to ten-fold molar excess of SMCC to keep the degree of labeling rather low (note that the molar excess used for organic dyes with an NHS ester is usually 15- to 20-fold). The unconjugated excess linker was separated by the antibody-SMCC complex by sephadex gel filtration (G25, PD10). The fractions containing the antibody molecules were determined using the BioRad Protein Assay (see subsection 5.4.1). Meanwhile, the FNDs had been treated with Traut's reagent (2-iminothiolane). Traut's reagent also reacts with primary amines (-NH$_2$) to introduce sulfhydryl (-SH) groups while maintaining charge properties similar to the original amino group. Finally, the SMCC-conjugated antibody was incubated with the thiol-decorated FNDs at pH 7.4 and 4 °C over night.

The tendency of the FNDs to form clusters, especially in aqueous environments, hampers the separation of FND-conjugated antibodies from free diamonds and leads, moreover, to large aggregates during the incubation on cells. Therefore, these agglomerations still need to be overcome to be a sufficiently applicable tool in the life sciences.

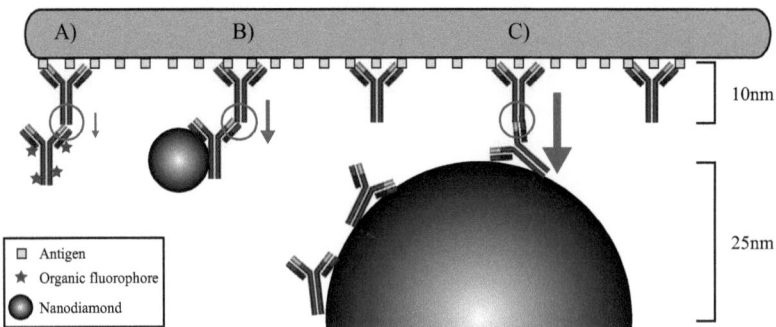

Figure 4.10: Model of the steric influence of FNDs on the functionality of antibodies. A) shows an example of standard indirect immunofluorescence labeling. A primary antibody binds to its antigen (yellow square). One or more secondary antibodies, conjugated to several organic fluorophores, bind to the F$_c$ terminus of the primary antibody. In **B)** the secondary antibody is conjugated to a FND of $\sim$ 10 nm in diameter, whereas the FND in **C)** has a diameter of $\sim$ 50 nm. The red arrows indicate the extra forces that exert on the non-covalent binding sites between primary and secondary antibodies (red circles).

5 Materials and Methods

5.1 Buffers and Solutions

For most of the experiments phosphate buffered, Tris or HEPES buffered solutions were used. All buffers and solutions were prepared with de-ionized water (ELGA LabWater, Celle, Germany) The pH was adjusted with a pH electrode (Basic pH Meter PB-11, Sartorius, Göttingen, Germany) using the accordant acids or bases respectively.

Buffer Name	Composition	pH
PBS / AB-PBS Phosphate Buffered Saline	137 mM NaCl, 2.68 mM KCl, 10 mM Na_2HPO_4, 1.76 mM KH_2PO_4	pH 7.4 / pH 6.5
TBS Tris Buffered Saline	10 mM Tris-(hydroxymethyl)- aminomethane/HCl, 150 mM NaCl	pH 7.5 - pH 9.0
HBS HEPES Buffered Saline	20 mM HEPES, 150 mM NaCl	pH 7.4
Fusion buffer	120 mM NaCl, 20 mM sodium acetate, 10 or 100 mM $CaCl_2$	pH 5.0
Liposome buffer	120 mM NaCl, 20 mM HEPES, 0.5 or 2 mM EGTA	pH 7.4
Sonication buffer	120 mM potassium glutamate, 10 mM EGTA, 20 mM potassium acetate, 4 mM $MgCl_2$, 2 mM ATP, 0.5 mM DTT, 20 mM HEPES-KOH	pH 7.2
PFA	4% or 10% (w/v) para formaldehyde in PBS	pH 7.4

Table 5.1: Composition of the used buffers and solutions

5.2 Lipids

For the study of lipid diffusion both labeled and unlabeled phosphoglycerolipids, sphingolipids and cholesterols were used (see 3.8). The conjugation of the lipids with a fluorescent dye molecule was done either by Atto-Tec (Siegen, Germany) or by Dr. Vladimir

Materials and Methods

Belov's group (MPI for biophysical Chemistry, Göttingen). To introduce labeled lipids into the plasma membrane different techniques were applied: 1. Liposomes containing labeled and unlabeled lipids and 2. BSA-conjugated labeled lipids.

5.2.1 Liposome generation

First, liposomes containing one kind of labeled lipid were made, meaning, that for every labeled lipid a set of liposomes had to be prepared. The unlabeled lyophilized lipids listed below were dissolved in chloroform:methanol (2:1) to a stock concentration of 25 mg/ml and stored at -20 °C under nitrogen gassing to avoid oxidation. The stock solution of the labeled lipids was 1 mg/ml.

Abbreviation	Systematic Name	MW [Da]
PE	1-octadecanoyl-2-(9Z-octadecenoyl)-*sn*-glycero-3-phosphoethanolamine	746.06
PC	1-hexadecanoyl-2-(9Z-octadecenoyl)*sn*-glycero-3-phosphocholine	760.09
PS	1-octadecanoyl-2-(9Z-octadecenoyl)-sn-glycero-3-phosphoserine	812.05
PI	1-octadecanoyl-2-(5Z,8Z,11Z,14Z-eicosatetraenoyl)-*sn*-glycero-3-phospho-(1'-myo-inositol)	909.12
C	Cholesterol	386.66
PE-Atto647N	Atto647N-1,2-dihexadecanoyl-*sn*-glycero-3-phosphoethanolamine	1492.06
SM-Atto647N	Atto647N-Sphingosylphosphorylcholine	1477.09
DOPE-Atto647N	Atto647N-1,2-dioctadecenoyl-*sn*-glycero-3-phosphoethanolamine	1490.06

Table 5.2: Labeled and unlabeled lipids used for liposome generation

Liposomes of 3500 nmol lipids were prepared by mixing the different lipids dissolved in chloroform/methanol at ratios as seen in table 5.3. When labeled PE lipids were added the amount of unlabeled PE was reduced by the amount of the labeled ones (650 nmol instead of 700 nmol). By introducing labeled SM into the liposome mixture it was added without reducing the amount of the other lipids, thus the liposomes were prepared using a 3550 nmol preparation mix. Then the mixture was evaporated under a stream of nitrogen until it was completely dried. Subsequently, it was redissolved in 500 μl calcium-free HEPES-EGTA buffer (see table 5.1). The mixture was vortexed and incubated at room temperature for one hour to allow swelling. Then it was sonicated in an ultrasonic bath at

4 °C for an additional hour. The lipid emulsion was then introduced into the LiposoFast extruder (Avestin, Ottawa, Canada) where unilamellar liposomes were generated by manual extrusion (50 cycles) through a polycarbonate membrane (pore diameter 50-100 nm) using two Hamilton glass syringes on each side of the membrane containing steel housing. The liposome emulsion was used up to two days when stored at 4 °C

Lipid labeled/ unlabeled	Percentage [%]	Amount of substance n [nmol]	Volume [μl] $V=\frac{M*n}{c}$
PE	20	700 or 650	20.9
PC	50	1750	53.2
PS	10	350	11.4
PI	10	350	12.7
C	10	350	5.4
PE-Atto647N	1.4	50	74.6
SM-Atto647N	1.4	50	73.9
DOPE-Atto647N	1.4	50	74.5

Table 5.3: Composition of liposomes

5.2.2 Generation of Plasma Membrane Sheets

After the incubation with the liposomes the cells were washed in serum-free DMEM (Dulbecco's Modified Eagle Medium) without phenolred to wash the non fused liposomes away. A technique to remove the upper plasma membrane (the one not opposing the coverslip) was applied, because of the rather high background due to the remaining non fused liposomes sticking to the cells. For this procedure the cells were transferred into a glass dish with a diameter of at least 10 cm filled with cold (4 °C) sonication buffer (composition see table 5.1) up to 3 to 4 cm in height. Then a sonication tip was placed directly over the center of the coverslip in a distance of 1 cm and one sonication pulse was applied (Branson Sonifier 450, G. Heinemann Ultraschall- und Labortechnik, Schwäbisch Gmünd, Germany). By varying the height of the sonication tip or by adjusting the pulse strength the percentage of membrane sheets could be optimized. Best results were obtained for a pulse strength of 3 to 3.5 and a pulse duration of 100 ms. These settings vary when different cell types are used from pulse strength of 1 for SH-SY5Y cells up to the above mentioned strength of 3.5 for PtK2 cells (see also discussion). The desired cells containing only the lower plasma membrane (or only small remains of the upper one) are

Materials and Methods

called membrane sheets and are still functional (Avery et al., 2000). To verify the membrane sheet generation cells were incubated with TMA-DPH, a lipid membrane marker (see also 5.5.1). Upon incorporation into lipid membranes its fluorescence is dramatically increased when excited in the UV range. Intact cells or cells consisting of only one plasma membrane can easily be distinguished by widefield fluorescence microscopy (see 5.6.1). The dynamic measurements, however, were done using STED-FCS (see 3.2.1) in a special microscope chamber (RC-40, Warner Instruments, Hamden, USA). Subsequent to the membrane sheet generation, they could be measured for up to 10 to 15 min.

5.2.3 Formation of the lipid-BSA complex

A complex formation between BSA and lipids facilitates their incorporation into the plasma membrane. The protocol introduced by Pagano and Martin (Pagano et al., 1989) was modified by Günter Schwarzmann (University of Bonn, Germany; private communication) and further modified in this thesis. The overall labeling can be controlled much simpler than by using liposomes and no generation of sheets is necessary to avoid background.

Introduction of Lipids According to Pagano and Martin

The stock solutions of the labeled lipids were also prepared in chloroform/ methanol to a concentration of 1 mg/ml under nitrogen gassing. 250 nmol lipid were dried under a stream of nitrogen and then redissolved in $200\,\mu l$ ethanol. The BSA-solution was prepared by dissolving 1.7 mg fatty acid-free BSA (AppliChem, Darmstadt, Germany) in 10 ml of HDMEM (phenol-free DMEM buffered with 10 mM HEPES). The redissolved lipids were slowly added to the freshly prepared BSA solution while vortexing to allow complex formation. A dialysis against 500 ml HDMEM at 4 °C overnight decreased the ethanol concentration. Then the lipid-BSA complexes could be aliquoted and stored at -20 °C.

For incubation with cells dilutions of the labeled lipid complexes (PE-Atto647N or SM-Atto647N) were prepared in HDMEM. The cells were kept on ice and incubated with the lipid-BSA complex for 30 min. Then cells were washed in cold HDMEM and incubated at 37 °C for 4 min in HDMEM (PE, PE-1, SM and CPE) before measurement. The gangliosides (GM1, GM1-# and GM1-##) and the ceramide, however, were directly prepared for measurement to prevent rapid internalization and, after brief washing in cold

HDMEM, mounted in the microscope chamber of the STED-FCS setup. All dynamic measurements were completed before severe morphological changes, as was checked by the cellular appearance in transmission light prior to each measurement, or internalization could occur at either 27 °C or 37 °C. The temperature was precisely controlled by the use of an objective heater (Bioptechs Inc., Butler, USA).

Introduction of Lipids According to Schwarzmann

In this protocol the concentration of BSA was increased compared to the method of Pagano and Martin. For complex formation 100 nmol lipid were dried under nitrogen gassing and moistened with 20 μl of absolute ethanol. Then 1 ml of a 7 mg/ml or 10^{-4} M (100 μM) concentrated BSA solution was added and the mixture heavily vortexed. After addition of 9 ml of HDMEM the final BSA concentration is 0.7 mg/ ml or 10^{-5} M (10 μM). Since the final ethanol concentration is 0.2% (v/v) no dialysis is required.

Another approach in which ethanol is completely left out is the direct dissolving of the dried lipids in the BSA-HDMEM solution at 37 °C for 30 min while vortexing. However, most dried lipids were poorly solubly.

For incorporation of the BSA-conjugated lipids into the plasma membrane, they were diluted in cold HDMEM (for concentrations please refer to table 5.4) and incubated on ice for 30 min. After a brief washing step in cold HDMEM the cells were directly transferred to the microscope chambers to either perform single molecule analysis, STED-FCS measurements, FRAP or back exchange experiments. However, some lipids (PE, PE-1, DOPE, SM and CPE) required prior to the washing an additional incubation step at 37 °C for 4 min to ensure a homogeneous distribution over the plasma membrane.

Verification of Lipid Incorporation with Back Exchange Experiments

Back exchange experiments were performed to verify the proper membrane incorporation of the fluorescently labeled lipids. The cells were incubated with the fluorescent lipid analogs of SM and GM1 as described before but with an approximately ten-times higher lipid concentration compared to the samples prepared for STED-FCS (see table 5.4). Briefly, cells were incubated with the BSA-lipid complexes diluted in HDMEM at 4 °C for 30 min. After an additional incubation step in HDMEM at 37 °C for 4 min and a brief washing step in in the case of the SM analogs cells were transferred to a microscope

Materials and Methods

Lipid	Number of Acylchains	Dye position	BSA conc. [mg/ml]	Single molecule/ FCS imaging
PE-Atto647N	2	head	1.4	10 nM-300 nM/ 1-3 μM
PE-1-Atto647N	(2)	replaces acyl chain	0.7	5 nM-50 nM/ 100-300 nM
CPE-Atto647N	2	head	1.4	$\sim$ 5 μM/ 5 μM-15 μM
SM-Atto647N	(2)	replaces acyl chain	0.7	5 nM-50 nM/ 300-500 nM
SM-NBD	(2)	replace acyl chain	0.7	$-$ / 300 nM-500 nM
SM-BODIPY	(2)	replaces acyl chain, C_6-linker	0.7	$-$ / 300 nM-500 nM
SM-Atto532	(2)	replaces acyl chain	0.7	$-$ / 300 nM-500 nM
DOPE-Atto647N	2	head	1.4	10 nM-300 nM/ 1 μM-3 μM
GM1-Atto647N	(2)	replaces acyl chain	0.7	5 nM-50 nM/ 200 nM-500 nM
GM1-NBD	(2)	replaces acyl chain, C_6-linker	0.7	$-$ / 200 nM-500 nM
GM1-BODIPY	(2)	replaces acyl chain, C_5-linker	0.7	$-$ / 200 nM-500 nM
GM1-#-Atto647N	2	head	1.4	$\sim$ 5 μM/ $\sim$ 10 μM
GM1-##-Atto647N	(3)	additional acyl chain	1.4	5 nM-50,nM/ 100 nM-300 nM
Cer-Atto647N	(2)	replaces acyl chain	0.7	$\sim$ 1 μM/ 2 μM-4 μM

Table 5.4: Lipids used for BSA complexes. The brackets indicate that one of the acyl chains is not native but partly substituted with the label.

chamber and imaged with a conventional laser scanning confocal microscope (see 5.6.2). The GM1 cells, however, could been imaged directly after the brief washing step. Thereupon, cells were treated with BSA-HDMEM (10^{-5} M BSA) on ice for 10 min and imaged again using the same settings as for the previous images. Fluorescent lipids, which were not incorporated properly, were removed from the membrane by BSA, resulting in a reduction of the fluorescence signal. Therefore, the fluorescence intensities of the images taken before and after the BSA treatment were compared to draw conclusions about the incorporation efficiency.

FRAP measurements

The FRAP experiments were performed on a confocal laser scanning microscope (TCS SP5, Leica Microsystems, Wetzlar, Germany) using the FRAP wizard included in the software (LAS AF, Leica Microsystems, Wetzlar, Germany). The fluorophore conjugated lipids (PE and SM) were incorporated in the plasma membrane. All images were recorded using a 63x high numerical oil lens (HCX PL APO lambda blue 63x NA 1.4 OIL UV) and a scanning speed of 400 lines per second. At first, an overview image was taken ($\sim 60 \times 60\,\mu m^2$, 1024 x 1024 pixels) with reduced laser intensity. Then, a region of interest (ROI) was marked ($\sim 15 \times 15\,\mu m^2$) and by zooming in, this ROI was rapidly photobleached within 15-20 images using a ten- to twenty-times higher laser intensity. This number of images was determined experimentally by observing, when the initial fluorescence had dropped to the background noise (usually after ten images). To ascertain that the fluorescent lipids in the ROI are bleached completely, the number of images was chosen to be slightly higher, 15 to 20. The recovery of the fluorescence was recorded with reduced laser intensity for ~ 20 frames and visualized graphically as a function of time.

Investigation of the Lipophilicity of the Fluorescent Lipids by Thin layer Chromatography

Analytical thin layer chromatography (TLC) of the gangliosides was performed on high performance thin layer chromatography (HPTLC) plates (Merck, Darmstadt, Germany) with silica gel 60 (eluent $CHCl_3$/ MeOH/ 15 mM $CaCl_2$, 60:35:8 (v/v)). The spots were then visualized by anisaldehyde reagent followed by heating at 150 °C for 5 min.

5.2.4 Cholesterol depletion

The endogenous cholesterol was depleted by either enzymatic oxidization by the oxidoreductase cholesterol oxidase (COase) or by removal via β-cyclodextrin (β-CD). A recovery of the cholesterol content in the cells was observed at after more than 30 min in the absence of the accordant reagents of depletion, leading to a restricted time window for the dynamics measurements. As the incubation with the labeled lipids was always carried out at 4 °C, thus not allowing for cholesterol recovery, it could be either performed before or after this treatment. However, it was more convenient to introduce the labeled lipids after

Materials and Methods

the cholesterol depletion to reduce their internalization.

Cholesterol Oxidase

The oxidoreductase cholesterol oxidase (COase) derived from *Streptomyces sp.* (Sigma Aldrich Chemie GmbH, Munich, Germany or Calbiochem, Merck, Darmstadt, Germany) catalyzes oxidation of cholesterol (K_M = 13.0 μM) to cholestenone and hydrogen peroxide. For cholesterol depletion from the plasma membrane with COase, a 1 mg/ml ($\sim$ 34 U/ml) stock solution was freshly prepared in 50 mM KH_2PO_4, pH 7.0. 60 μl of this stock was further diluted in 2 ml of prewarmed HDMEM (without phenol red) resulting in a final concentration of c = 1 U/ml. The cells seeded one day prior on coverslips were incubated with the COase containing HDMEM at 37 °C for 30 min. Followed by a washing step in prewarmed HDMEM cells were (unless already done before) incubated with the labeled lipids as described above (see 5.2.3).

β-Cyclodextrin

Cyclodextrins are cyclic oligosaccharides which are able of forming host-guest complexes with hydrophobic molecules such as cholesterol. Therefore it can be used as tool to remove cholesterol from the plasma membrane. The cholesterol depletion with β-CD was performed using a 10 mM solution prepared in HDMEM. The cells were incubated at 37 °C for 45 min. Incubation with the labeled lipids was, according to the COase treatment, carried out after washing in warm HDMEM in the usual way at 4 °C.

5.3 Organisms and organism specific methods

5.3.1 Cultivation of *Escherichia coli*

For amplification and purification of plasmid DNA the *E. coli* strain DH5α was used. The cultivation as liquid overnight cultures was carried out in LB-medium (Luria Bertani medium: 0.5% (w/v) yeast extract; 1% (w/v) pepton; 0.5% (v/v) 1 N NaOH; 0.5% (w/v) NaCl) and on agar plates (LB medium supplemented with 2% (w/v) agar) respectively. For selection the antibiotics ampicillin (50 μg/ml; Sigma Aldrich Chemie GmbH, Munich, Germany) or kanamycin (50 μg/ml; Applichem, Darmstadt, Germany) were added.

For permanent cultures 500 µl of a liquid overnight culture was mixed with 500 µl of 80% (v/v) sterile glycerol in cryotubes. After flash freezing in liquid nitrogen they were stored at -80 °C.

5.3.2 Transformation of *E. coli*

Transformation of *E. coli* via electroporation requires electrocompetent bacteria. According to Dower *et al.* an *E. coli* culture was grown till the logarithmic growth phase ($OD_{600} \sim 0.6$) and incubated on ice for 30 minutes. Afterwards the cells were pelleted by 4000 g at 4 °C for 15 minutes (Sorvall RC-5B, DuPont Instruments, Wilmington, DE, USA). This was followed by a two minutes washing step in ice cold water and another washing step in 1/50 volume 10% (w/v) glycerol. Finally the cells were resuspended in 1 ml of 10% (w/v) glycerol, aliquoted (40 µl), flash frozen in liquid nitrogen and then stored at -80 °C.

For electroporation the cells were thawed on ice. Then 20 ng-40 ng plasmid DNA were added, and incubated on ice for five minutes and finally transferred into an electroporation cuvette (diameter 2 mm). The electroporator was set to 2.5 kV, 200 Ω and 25 µF and a brief voltage pulse applied. The cells were then transferred into a fresh reaction tube and incubated in LB medium at 37 °C for one hour while shaking. Afterwards they were plated on agar plates containing antibiotics for selection and incubated at 37 °C overnight.

5.3.3 Isolation and purification of plasmid DNA from *E. coli*

The isolation of plasmid DNA from *E. coli* is based on alkaline lysis of the bacterial cell wall ((Birnboim and Doly, 1979)). After enzymatic digestion of cellular RNA and genomic DNA the plasmid DNA can be purified. Here the EndoFree Plasmid Maxi Kit (Qiagen, Hilden, Germany) was used following the manufacturer's instructions. Briefly, for a large scale preparation 200 ml of LB medium supplemented with an antibiotic were inoculated and grown overnight at 37 °C while shaking. The following day, the cells were pelleted at 6000 g for 15 min (Sorvall RC-5B, DuPont Instruments, Wilmington, DE, USA) and resuspended in 10 ml of lysis buffer (Qiagen buffer P1: 50 mM Tris/ HCl; 10 mM EDTA; 100 µg/ml RNase A; pH 8.0). After addition of 10 ml of the alkaline SDS solution (Qiagen buffer P2: 200 mM NaOH; 1% (w/v) SDS) a five minutes incubation

Materials and Methods

with 10 ml of neutralization buffer (Qiagen buffer P3: 3 M potassium acetate; pH 5.5) was following and the whole suspension was then transferred into a catridge. After a 15-minute incubation at room temperature the proteins precipitated on the surface were removed by filtration through the catridge, whereas the clear solution contained the DNA. The clear solution was supplemented with 2.5 ml of endotoxin removal buffer (Qiagen buffer ER, contains propan-2-ol and polyethylene glycol octylphenyl ether) mixed and incubated on ice for 30 minutes. Then the lysate was transferred onto an equilibrated column (Qiagen buffer QBT) resulting in the binding of the plasmid DNA to the matrix. After two washes (Qiagen buffer QB) the plasmid DNA was eluted with 15 ml (Qiagen buffer QN) into a centrifuge tube. The additon of 10.7 ml isopropanol followed by an immediate centrifugation at 4 °C for 30 minutes (Sorvall RC-6 Plus, Thermo, Thermo Fisher Scientific, Waltham, MA, USA) resulted in the precipitation of the plasmid DNA. The pellet was washed in 70% ethanol and air dried. Finally the plasmid DNA was resuspended in an appropriate volume of de-ionized water.

5.3.4 Mammalian cell lines

In this work different mammalian cell lines were used. For studying the lipid diffusion the adherent, epithelial cell line **PtK2** (Australian long-nose potoroo *Potorous tridactylus*, kidney) was chosen for its wide spread and flat growth (Courtesy of Prof. Dr. Mary Osborne).

The cell line **U373MG** is a human glioblastoma astrocytoma derived from a malignant tumor by explant technique (Courtesy of). Unlike PtK2 cells they don't form monolayers.

For the study of neurofilaments the human neuroblastoma cell line **SH-SY5Y** was used (DSMZ, Braunschweig, Germany). This cell line was thrice subcloned from the original metastatic bone tumor SK-N-SH of a four year old girl. SH-SY5Y consists of three morphological different cell types (S-type, N-type, I-type). The proportion of these types can be changed by special treatments (see subsection 5.3.6).

5.3.5 Media and cultivation

The cultivation of all the cell lines was done at 37 °C with 5% CO_2 in a water saturated atmosphere. For all the following media the basic ingredients such as DMEM (Dulbecco's Modified Eagle Medium) and RPMI 1640 (Roswell Park Memorial Institute) supplemented with either GlutaMAX or L-glutamine and various glucose concentrations were purchased from Invitrogen (Carlsbad, CA, USA). Penicillin was added to a final concentration of 20 U/ml and streptomycin to a final concentration of 20 μg/ml, respectively (Biochrom AG, Berlin, Germany). Additional supplements such as FBS (Fetal Bovine Serum, Invitrogen) and Sodium pyruvate (Sigma Aldrich Chemie GmbH, Munich, Germany) were added as seen below in the table. All media were sterile filtrated.

Cell type	Medium	Serum	Glucose	Pyruvate	Supplements
PtK2 cultivation	DMEM	10% FBS	4.5 g/l	1 mM	no
U373MG cultivation	RPMI	10% FBS	4.5 g/l	1 mM	no
SH-SY5Y + cultivation	DMEM	10% FBS	4.5 g/l	1 mM	no
SH-SY5Y - cultivation	DMEM	10% FBS	1.0 g/l	1 mM	no
SH-SY5Y differentiation I	DMEM	10% FBS	4.5 g/l 1.0 g/l	1 mM	10 μM RA
SH-SY5Y differentiation II	DMEM	no	4.5 g/l 1.0 g/l	1 mM	50 ng/ml hBDNF
SH-SY5Y deprivation A	DMEM	no	no	1 mM	50 ng/ml hBDNF
SH-SY5Y deprivation B	DMEM	no	4.5 g/l	no	50 ng/ml hBDNF
SH-SY5Y deprivation C	DMEM	no	no	5.6 mM	50 ng/ml hBDNF

Table 5.5: The different media used for the cultivation, differentiation and deprivation experiments. The + and - signs indicate whether high or low glucose was used. Moreover, two different differentiation media (I and II) and three different deprivation media (A, B, C) varying in their glucose and pyruvate contents were prepared

All cells were grown in T25 flasks (Sarstedt, Nürnbrecht, Germany) and splitted in a five to seven days rhythm using trypsin/ EDTA (0.05%/ 0.02% in PBS) to detach the cells from the surface. They then were seeded in a ratio of 1:10 to 1:20 for SH-SY5Y and U373MG and 1:30 to 1:40 for PtK2 in fresh cultivation medium. For biochemical experiments cells were seeded in 10 cm diameter petri dishes (Sarstedt, Nürnbrecht, Germany)

Materials and Methods

whereas for microscopy studies cells were seeded on coverslips (Menzel, Braunschweig, Germany) and grown to 80% confluence.

Cryopreservation for long-term storage was used for all the cells. Subconfluent cells were trypsinized and after weak centrifugation resuspended in freezing medium (50% (v/v) cultivation medium, 40% (v/v) FBS, 10% (v/v) DMSO) and transferred into cryotubes (Nunc, Thermo Fisher Scientific, Berlin, Germany). A slow, constant cooling rate of 1 °C per minute was achieved by using a freezing container, Mr. Frosty (Nalgene, Thermo Fisher Scientific, Rochester, NY, USA). For long-term storage the cryotubes were then transferred into liquid nitrogen.

5.3.6 Re-differentiation of SH-SY5Y

The differentiation was usually carried out using cells seeded on coverslips. A stock solution of 10 mM all-*trans* retinoic acid, RA (Calbiochem, Merck, Darmstadt, Germany) was prepared in DMSO and then directly added to the culture medium resulting in a final concentration of 10 μM (see SH-SY5Y differentiation medium I in table 5.5). After five days in the presence of RA, cells were washed three times with serum free DMEM and then incubated with differentiation medium II 5.5), containing no serum but 50 ng/ ml of hBDNF (human brain derived neurotrophic factor; concentration of stock solution: 10 ng/ μl in water). With replacing the medium every five days, the cells could be kept for several weeks in the presence of hBDNF.

5.3.7 Test on mycoplasma

For testing of mycoplasma contamination two methods were used. Staining with DAPI (4',6-Diamidino-2-phenylindole, Roche, Mannheim, Germany) simply relies on the detection of total DNA present in the culture dish. The cells were seeded on coverslips, PFA or methanol fixed and incubated in a solution containing 2.5 mg/ ml DAPI for five minutes. The fluorescence was detected and analyzed by fluorescence microscopy. Uncontaminated cells show only nuclear fluorescence against a dark cytoplasmic background (sometimes a weak mitochondria staining can be observed due to the binding of DAPI to mitochondrial DNA). Infected cells however, are detected as bright foci in the cytoplasm and sometimes also in the intracellular space. This method is quick and easy but can only detect severe contaminations.

By using the 'Lookout Mycoplasma PCR Detection Kit' (Sigma Aldrich Chemie GmbH, Munich, Germany) according to the manufacturer's protocol a contamination can be detected much more precisely. This kit is based on PCR and detects the highly conserved 16S rRNA coding region of the mycoplasma genome. The kit provides the primer set and a DNA control sample. 100 μl of cell culture supernatant of cells grown to 90-100% confluence are boiled at 95 °C for 5 min and centrifuged before adding to the PCR mixture. Total volume for each PCR is 25 μl. Additionally the included positive control and one negative control should always be performed. After agarose gel separation of the PCR a band at 481 bp indicates an infection of mycoplasma.

5.3.8 Transfection of mammalian cells

To introduce the plasmid DNA encoding for the GPI anchor fused to avidin (kindly provided by Fabien Pinaud), cells were seeded one day prior to the transfection on coverslips. Here, nanofection with Nanofectin (PAA, Pasching, Austria) was used according to the manufacturer's protocol. This technique is based on the binding of DNA molecules to a positively charged polymer that protects the DNA from degradation by proteases. Since this reagent is not toxic, no medium change was required. Furthermore, it can be used in the absence or presence of serum. In brief, 3 μg DNA was used for the transfection of one well of a 6well plate (3 ml) and mixed with the polymer-containing solution. After a short incubation step at room temperature for 15 min to 30 min, the DNA-polymer complex was added drop-wise to the cells. The expression of the fusion protein could typically be observed after 24 h.

5.4 Protein biochemistry

To study composition and modifications of proteins biochemically, standard methods such as Western immunoblotting and enzyme assays were used.

Materials and Methods

5.4.1 Western analysis

Western Blotting, Western blot or immunoblotting allows to determine, with a specific primary antibody, the relative amounts of a specific protein present in different samples. Briefly, samples are prepared from cells that are homogenized in a buffer that protects the protein of interest from degradation. The sample is separated using SDS-PAGE and then transferred to a membrane for immunodetection. With the conversion of a chemiluminescent developing solution by the antibody-enzyme conjugate the protein of interest can be highlighted.

Isolation of total cell protein from mammalian cells

The cells were seeded in a 10 cm diameter petri dish and grown to about 80% confluence. After aspiration of the culture medium cells were washed two times with PBS. As the extracted proteins were used in downstream applications, such as enzyme assays (see 5.4.2), a reagent avoiding protein degradation or activity interference was used for cell lysis (CelLytic M Cell Lysis Reagent; Sigma Aldrich Chemie GmbH, Munich, Germany). For Western analysis a protease inhibitor cocktail was added to this reagent (Complete Protease Inhibitor Cocktail; Roche, Mannheim, Germany). 500 μl of lysis buffer were pipetted directly on the PBS washed cells and incubated for 15 to 30 min at room temperature while shaking gently.

Protein content

The protein content was determined using the BioRad Protein Assay (BioRad, Hercules, CA, USA). This method is based on the Bradford assay (Bradford, 1976) which relies on the complex formation of proteins with a phosphoric acid containing dye solution. According to the manufacturer's protocol 5 μg of the protein samples were mixed with the dye solution (1:5 in water). After incubation for five minutes the absorbance was read at 595 nm. By comparing the value with a calibration curve the protein concentration was determined.

5.4.1 Western analysis

SDS-PAGE

The protein samples were separated according to their size using the discontinuous sodium dodecyl sulfate polyacrylamide gel electrophoresis (SDS-PAGE) (Laemmli, 1970). Dependent on the molecular weight of the proteins of interest, the acrylamide concentration in the separation gel varied between 8 and 15% (see table 5.6). The lower the acrylamide concentration the better the separation in the high molecular weight range.

For the analysis of the neurofilament proteins (MW 68kD - 200kD) an 8% separation gel was used, whereas for the enzyme assays a 12.5% gel was used.

	Separation Gel 8%	Separation Gel 12.5%	Stacking Gel 5%
Water	6.95 ml	4.7 ml	5.6 ml
1.5 M Tris/HCl pH 8.8	3.75 ml	3.75 ml	-
0.5 M Tris/HCl pH 6.8	-	-	2.5 ml
Acrylamide 30% bisacrylamide solution	4 ml	6.25 ml	1.7 ml
SDS 10% (w/v)	150 μl	150 μl	100 μl
APS 10% (w/v)	150 μl	150 μl	100 μl
TEMED	15 μl	15 μl	10 μl

Table 5.6: Composition of the different separation gels and the stacking gel

5x Laemmli Sample Buffer (2% (w/v) SDS, 0.8 M βmercaptoethanol, 10% (v/v) glycerol, 0.2% bromphenol blue, 62.5 mM Tris/ HCl pH 6.8) was added to the protein extracts. To achieve complete solubilizing it was then heated up at 95 °C for five minutes. After loading (15 to 30 μg) the samples were electrophoretically separated at 20 to 30 mAc in the presence of electrophoresis buffer (see table 5.7).

As protein size marker a prestained protein ladder (Spectra Protein Ladder, MBI Fermentas, Burlington, Canada) was loaded.

Protein transfer

To expose the proteins in a way that they can be detected immunologically they need to be transferred from the gel to a membrane. Here polyvinylidene fluoride membranes (PVDF, Hybond-P, GE Healthcare, Freiburg, Germany) with a pore size of 0.2 μm were used. The transfer was performed using standard tank electroblotting: the membrane was placed directly on the gel and sandwiched by several layers of Whatman papers and two

Materials and Methods

sponges soaked with transfer buffer (see table 5.7). The transfer was done at 4 °C either for 2.5 h at 400 mA/ 100 V or overnight at 80 mA/ 25 V in transfer buffer.

Buffer Name	Composition	pH
Electrophoresis buffer	25 mM Tris/ HCl, 192 mM glycine, 0.1% (w/v) SDS	pH 8.3
Transfer buffer	20 mM Tris/ HCl, 150 mM glycine, 20% methanol (v/v)	pH 8.3
Blocking buffer (for Western Blot)	5% (w/v) non-fat dry milk, 0.05% (v/v) Tween20 in PBS	pH 7.4
Washing buffer (for Western Blot)	1% (w/v) non-fat dry milk, 0.05% (v/v) Tween20 in PBS	pH 7.4

Table 5.7: Composition of the used buffers and solutions

Ponceau S staining

To analyze the efficiency of protein transfer on the membrane a staining in Ponceau S solution (0.3% (w/v) Ponceau S, 3% (w/v) TCA in water) was carried out (Salinovich and Montelaro, 1986). The incubaton with this diazo dye for five to ten minutes allows for a reversible protein band labeling which after documentation can be easily removed by washing in destilled water.

Immunological detection of proteins

Prior to the incubation with the primary antibody the membrane was blocked for 45 min in blocking buffer (see table 5.7) to prevent unspecific binding of the antibody. Incubation with the primary antibody diluted in washing buffer (see 5.7) to a final concentration between 1 and 5 μg/ ml was carried out at room temperature for 1 h or at 4 °C overnight. After several washing steps the membrane was incubated with the HRP (Horseradish Peroxidase)-conjugated secondary antibody at room temperature for 1 h.

For the detection of the protein bands a chemiluminescent agent, luminol, in the presence of hydrogen peroxide (Western Lightning, PerkinElmer LAS GmbH, Rodgau, Germany) was used which produced luminescence after enzymatic reaction proportional to the amount of protein. This chemiluminescence was detected and documented by a luminescence detector (Lumi-Imager, Roche, Mannheim, Germany).

5.4.2 Kinase assays

In a kinase assay the activity of specific kinases can be determined by detecting their activated, phosphorylated form in an ELISA based assay or by measuring of their substrates' phosphorylation in an immunoprecipation based assay.

p38 MAP Kinase assay

The activity of p38 MAPK was determined using a nonradioactive assay based on immunoprecipitation of the active kinase. The kit was used according to the manufacturer's protocol (p38 MAP Kinase Assay Kit, Cell Signaling Technology, Danvers, MA, USA). Cells were grown under different conditions in 10 cm diameter petri dishes. To harvest the cells under non denaturing conditions after aspiration of the medium either 500 μl of the provided Cell Lysis Buffer (20 mM Tris, 150 mM NaCl, 1 mM EDTA, 1 mM EGTA, 1% Triton, 2.5 mM sodium pyrophosphate, 1 mM β-Glycerolphosphate, 1 mM Na_3VO_4, 1 μg/ml Leupeptin, pH 7.5) supplemented with 1 mM PMSF (phenylmethylsulfonyl fluoride) or 500 μl of the above mentioned CelLytic M Cell Lysis Reagent (see 5.4.1) were pipetted directly on the cells. After an incubation on ice for five minutes (provided Lysis Buffer) or at room temperature for 15 minutes (CelLytic M) the cells were scraped off and transferred into appropriate tubes. Then sonication on ice was carried out four times five seconds each followed by a centrifugation at 4 °C for ten minutes. 200 μl of the supernatant was then incubated with 20 μl of the immobilized Phospho-p38 MAPK (Thr180/182) monoclonal mouse IgG at 4 °C with gentle rocking overnight such binding the phosphorylated kinase to the agarose beads. After microcentrifugation at 14,000g for 60 seconds at 4 °C the immunoprecipitated active kinase was in the pellet which then was washed twice with 500 μl of lysis buffer followed by two washing steps with 500 μl of kinase buffer (25 mM Tris, 5 mM β-Glycerolphosphate, 2 mM DTT, 0.1 mM Na_3VO_4, 10 mM $MgCl_2$, pH 7.5). Afterwards the pellet is suspended in 50 μl of kinase buffer supplemented with 200 μM ATP and 1 μg of kinase substrate (ATF-2 fusion protein) and incubated ar 30 °C for 30 minutes. The reaction is terminated by the application of 25 μl of 3x SDS sample buffer, vortexed and microcentrifuged at 14,000g for 60 seconds.

For reading the detailed procedure of SDS-PAGE and immunoblotting please refer to 5.4.1. Briefly, on a 12.5% SDS-gel 30 μl of the samples were loaded and afterwards transferred on a PVDF membrane. The primary antibody anti-Phospho-ATF-2 (Thr76)

Materials and Methods

rabbit IgG was diluted 1:1000 in washing buffer (see 5.1) at 4 °C overnight followed after several washes by the incbation with the HRP-comnjugated secondary antibody. The detection was done with the Lumi-Imager by using the provided chemiluminescent LumiGlo and peroxide according to the protocol.

JNK assay

To determine the activity of the stress-activated protein kinase/ c-Jun N-terminal kinase (SAPK/JNK) two assays were performed: one immunoprecipitation based assay (SAPK/JNK Kinase Assay Kit, Cell Signaling Technology, Danvers, MA, USA) similar to the one used for p38 MAPK and a second ELISA-based assay (Cellular Activation of Signaling ELISA CASE^{TM} Kit for JNK T183/Y185 Cellular Activation of Signaling ELISA, SABiosciences, Frederick, MD, USA). The IP-based kit required the preparation of cell lysates, which were prepared as described above (refer to the above chapter 'p38 MAPK assay'). In brief, a c-Jun fusion protein linked to agarose beads is used to pull down SAPK/JNK enzyme from the cell extracts. Upon addition of kinase buffer in the presence of ATP, SAPK/JNK phosphorylates the c-Jun substrate. The phosphorylated substrate was then detected by immunoblotting (Anti-c-Jun (phospho S63) rabbit IgG [Y172]) and can be used to determine the kinase activity.

The second ELISA-based method does not require cell lysates, but can be performed directly on cells seeded in 96-well plates. The kit for JNK is designed to analyze JNK phosphorylation in cultured human and mouse cell lines. This cell-based ELISA kit quantifies the amount of activated (phosphorylated) JNK protein relative to total JNK protein. The JNK phosphorylation sites are threonine 183 and tyrosine 185. In brief, cells grown in a 96-well plate are fixed with 4% (w/v) formaldehyde for 20 min, followed by a washing step and an incubation with quenching buffer, containing H_2O_2 and NaN_3, for 5 min. After the addition of antigen retrieval buffer the plate is placed in a microwave and heated at 375 W for 3 min. The antigen retrieval buffer is removed by washing, directly followed ba a blocking step for 1 h. Both antibodies, anti-pan-JNK and anti-phospho-JNK mouse IgG are added to each apropriate well and incubated for 1 h at room temperature. After washing, the secondary antibody is also incubated for 1 h at room temperature. After two washing steps, developing solution is added and incubated until the darkest staining well turns medium- to dark-blue. Then, stop solution is added and a color change to yellow can be observed. Within 5 min, the absorbance at 450 nm on an ELISA plate reader is read.

As JNK is known to be upregulated upon UV irradiation, cells were exposed to UV light for 10 sec as positive control. In the diagram (see 3.6) the relative OD values are shown for total and phosphorylated JNK. The cells were grown in the presence and absence of glucose for 15-24 h. Always two identical plates were prepared to irradiate one with UV light.

To normalize these obtained values to the cell number, cell staining buffer is added to the wells and incubated for 30 min at room temperature. After washing, 1% SDS is added and the plate incubated while gently rocking for 1 h. Finally, the OD is determined at 595 nm

5.5 Specific labeling

To visualize cellular structures they have to be specifically labeled either indirectly via antibodies or directly *in vitro*. Some structures and organelles such as the nucleus or mitochondria can be specifically tracked by certain dyes due to their chemical structure.

5.5.1 DNA dyes, TMA-DPH and mito trackers

For the detection of mycoplasma or staining of the nucleus in immunofluorescence the DNA intercalating dye DAPI (4',6-diamidino-2-phenylindole, Sigma Aldrich Chemie GmbH, Munich, Germany) can be used both in live and fixed cells due to its ability to pass also intact membranes. The typical concentration used was 2.5 μg/ml. The absorption maximum of DAPI bound to DNA is at 358 nm and the emission maximum is at 461 nm, in the blue range. DAPI also has the ability to intercalate in RNA but with a five times lower emission and an emission peak shifted slightly to the red (500 nm). The staining with DAPI can be done simply by adding it to the mounting medium for fixed cells or by adding it directly into the medium of living cells. Another DNA marker dye used in this work, SYTOX Orange (Molecular Probes, Eugene, OR, USA) was used in a concentration of 0.1 μM for ten minutes and has its emission peak at 570 nm when bound to DNA. This dye can only penetrate compromised membranes.

Materials and Methods

TMA-DPH (N,N,N-Trimethyl-4-(6-phenyl-1,3,5-hexatrien-1-yl)phenylammonium p-toluenesulfonate, Sigma Aldrich Chemie GmbH, Munich, Germany) is a lipophilic dye which interacts with lipid membranes. Its cationic trimethylammonium substitute acts as a surface anchor to improve the localization of the fluorescent probe of membrane interiors, DPH. It is used to verify sheet generation (see also 5.2.1). Unfortunately TMA-DPH is rapidly photobleached but, however, upon integration into the membrane strong enhancement of the fluorescence is observed. On the other hand the fluorescence in water is nearly negligible thus an excess of staining solution can be added. For staining the membranes a 1:20 dilution in sonication buffer is prepared from the saturated stock solution (in PBS).

As a viability marker for live cell observations two tracker dyes were used. The green fluorescent dye $DiOC_6(3)$ (3,3'-dihexyloxacarbocyanine iodide, Molecular Probes, Eugene, OR, USA) interacts with endomembranes like endoplasmic reticulum, mitochondria and vesicles. When used at very low concentrations as 0.1 nM in ethanol directly added to the culture medium or the microscopy medium after five to ten minutes only the mitochondria are stained. The other dye used was the cationic dye 5,5',6,6'-tetrachloro-1,1',3,3'-tetraethylbenzimidazolylcarbocyanine iodide (JC-1; $CBIC_2(3)$, Invitrogen, Carlsbad, CA, USA). JC-1 exhibits two states, a monomeric (green) and J-aggregates (orange-red). They can be used as an indicator for the mitochondrial potential, since the shift of the fluorescence from $\sim$525 nm to $\sim$590 nm is potential-sensitive. Thus, the depolarization state, and viability of the cells can be obtained by determining the red to green fluorescence ratio. The most widely implemented application of JC-1 is for detection of mitochondrial depolarization occurring in the early stages of apoptosis. The stock solution was prepared at 5 mg/ml in DMSO, the staining was carried out by directly adding it to the culture medium at $\sim$ 1 μg/ml and incubating it at 37 °C for 30 min. Polarized mitochondria are marked by punctuated red fluorescence (J-aggregate) which is replaced upon depolarization with 2.5 μM oligomycin for 30 min by diffuse green monomer fluorescence. Some of the green fluorescence remains associated with the mitochondria. Therefore, always both channels were recorded for control and deprived cells.

5.5.2 Dye Conjugation of proteins

For most of the dyes used for STED microscopy were non standard dyes there were no commercial antibody-dye conjugates ór other labeled proteins available. The conjugation was done with functionalized dyes and unmodified, BSA-free secondary, primary or recombinant proteins (see table 5.8). The principle of the conjugation relies on the reaction between the NHS-esters of the dye with the free amino groups of the protein in aqueous solutions. In general, a 15- to 20-fold molar excess of dye was used for all conjugations.

To a 1 mg/ml protein solution (usually dissolved in PBS) 1/10 volume of 1 M sodium bicarbonate solution (pH 8.5) was added to obtain a resulting pH of 8.3. 50 to 250 μg for primary and recombinant proteins, respectively, and 1 to 2 mg for secondary antibodies were used. The dye was dissolved in anhydrous DMF to a final concentration of 10 mg/ml and added slowly to the protein solution while stirring. After stirring for 1 h at room temperature 2 to 20 μl of a 1.5 M hydroxylamine solution were added to stop the reaction by quenching the remaining free NHS-esters. The separation of the protein-dye conjugate from the free dye was carried out using a Sephadex G-25 gel filtration column (PD-10 from GE Healthcare, Freiburg, Germany) pre-equilibrated with PBS pH 6.5. The first colored and fluorescent zone to elute is the conjugate whereas the second, slower moving zone contains the hydrolyzed free NHS-dyes. After determination of the protein content with the Bradford assay (see 5.4.1) the fractions were aliquoted and BSA was added to a final concentration of 5 mg/ml for long term storage at -80 °C.

Degree of labeling

The degree of labeling (DOL, dye-to-protein ratio) can be determined by absorption spectroscopy (according to the protocol provided by Atto-Tec, Siegen, Germany). According to the dependency stated in the Lambert-Beer law, the absorbance (A) is proportional to the extinction coefficient (ϵ), the molar concentration (c) and the optical path length (d):

A $= \epsilon * c * d$.

By simply measuring the UV-VIS spectrum of the conjugate solution the absorbance (A_{max}) at the absorption maximum (λ_{abs}) of the dye and the absorbance (A_{280}) at 280 nm (absorption maximum of proteins) can be determined. The concentration of conjugated dye is given by $c(dye) = \frac{A_{max}}{\epsilon_{max}*d}$ and the protein concentration can be calculated in the same way from its absorbance at 280 nm. (A_{280}) has to be corrected because the dyes also

Materials and Methods

Proteins	Company	Dye
Sheep anti-mouse IgG	Jackson ImmunoResearch Europe Ltd., Suffolk, UK	Atto590-NHS Atto647N-NHS KK114-NHS
Goat anti-rabbit IgG	Jackson ImmunoResearch Europe Ltd., Suffolk, UK	Atto590-NHS Atto647N-NHS KK114-NHS
Anti-vimentin [V9] mouse IgG	Sigma Aldrich Chemie GmbH, Munich, Germany	Atto590-NHS
Anti-neurofilament H phosph. [SMI36] mouse IgG	Abcam, Cambridge, UK	KK114-NHS
Anti-neurofilament M+H phosph. [SMI31] mouse IgG	Abcam, Cambridge, UK	Atto590-NHS
Neurofilament H recombinant protein	Progen, Heidelberg, Germany	Atto647N-Maleimide
Neurofilament L recombinant protein	Progen, Heidelberg, Germany	Atto647N-Maleimide

Table 5.8: Proteins and their conjugates

contribute to this value. This correction can be done by introducing a correction factor (CF_{280}) which is given by the quotient of $\frac{\epsilon_{280}}{\epsilon_{max}}$ such resulting in $A_{dye, 280} = A_{max} * CF_{280}$. The extinction coefficients are dye specific and were provided of the company (Atto dyes: Atto-Tec) or determined by the mannufacturers (KK114, Dr. V. Belov's group).

The absorbance for the proteins is now given by $A_{prot} = A_{280} - A_{max} * CF_{280}$ which leads to the protein concentration $c(prot) = \frac{A_{prot}}{\epsilon_{prot}}$ with ϵ_{prot} is the extinction coefficient of the protein at 280 nm. For the DOL follows then:

$$DOL = \frac{c(dye)}{c(prot)} = \frac{A_{max}/\epsilon_{max}}{A_{prot}/\epsilon_{prot}} = \frac{A_{max}*\epsilon_{prot}}{(A_{280}-A_{max}*CF_{280})*\epsilon_{max}}$$

If ϵ_{max}, the extinction coefficient of the free dye at the absorption maximum, is changed when the dye is bound to proteins it could lead to an error of up to 20%.

The DOL's for the used proteins varied between seven to twelve dye molecules per antibody or five to seven dye molecules for the recombinant proteins. In the latter case the heterogeneous labeling caused problems in the in vitro polymerization of the neurofilaments.

To overcome these random labelings maleimide-functionalized dyes which react with reduced thiols were used. For there are only few reduced cysteines in the neurofilament proteins (see primary sequences of NFH in the supplementary) a second labeling approach was done using Atto647N-maleimide. The conjugation was carried out as described for the NHS-ester with slight changes: the stirring time was 2 h at room temperature and instead of hydroxylamine glutathion or β-mercaptoethanol were used to ensure that no unconjugated dye is left.

5.5.3 Immunofluorescence

The detection of antigens by antibodies which are labeled with fluorescent dyes is called immunofluorescence. Most commonly the indirect immunofluorescence technique is used: A primary unlabeled antibody which recognizes the antigen and a secondary labeled antibody which binds to the constant organism specific F_c terminus of the primary antibody. That results in an amplification of the signal for several labeled secondary antibodies can bind to one primary antibody. The other technique also used here is the direct immunofluorescence. Hereby the primary antibody is directly labeled with the fluorophore resulting in a weaker signal but also in a decreasing of the antigen-antibody complex. This decrease is especially important when the resolution of the microscope reaches the size of these complexes, as in STED microscopy.

Besides antibodies also labeled toxins, e.g. phalloidin, can be used to highly specifically label cellular structures. Usually subconfluent cells (about 80% of confluence) were used for immunofluorescence and aside from the differentiated cells seeded 24 to 48 hours before on sterilized coverslips. The basic steps for immunofluorescence are fixation, extraction, detection and mounting.

Fixation

This step is crucial and dependent on the nature of the antigen and on the used antibody, so one has to be careful by choosing a fixative. Also it has to be taken into account that several epitopes can be masked upon fixation leading to a loss of antibody binding. There are two classes of fixatives: organic solvents and cross-linking reagents. Organic solvents such as alcohols and acetone remove lipids and dehydrate cells, while precipitating proteins. Whereas cross-linking agents such as aldehydes form intermolecular bridges. In general, with cross-linkers the cellular structure is better preserved but the antigenicity is reduced. Thus for every primary antibody both fixation techniques were tested.

Methanol fixation is quick and easy and no additional extraction step is necessary. For membrane enclosed organelles such as mitochondria or the Golgi apparatus and mobile cytosolic proteins para-formaldehyde is the fixative of choice. Whether the antigen is

Materials and Methods

masked by the fixation or not had to be tested for all used antibodies. For most of the cytoskeletal proteins (microtubules and all used intermediate filaments) methanol (-20 °C) fixation for 4 min worked best.

For labeling the F-actin stress fibers with phallotoxins the cells had to be fixed with 4% (w/v) para-formaldehyde in PBS for 15 min.

Extraction and blocking

The fixation with organic solvents also permeabilizes the plasma membrane so no further treatment is necessary. Upon fixation with para-formaldehyde (or other cross-linking reagents) a five to ten minutes extraction step with 0.1% (v/v) Triton X-100 in PBS is required. Before incubating the cells with the primary antibody (whether conugated to a fluorophore or unconjugated) unwanted binding sites have to be masked to decrease unspecific background. The cells were therefore incubated in blocking buffer 1-5% (w/v) BSA in PBS for 10 minutes.

Detection

After the blocking step the primary antibody was diluted to a final concentration of 1-5 μg/ml in 1% (w/v) BSA in PBS and incubated in a humid chamber either at room temperature for 1h or at 4 °C overnight. This incubation was followed by several washes in 1% (w/v) BSA in PBS. In the case of direct immunolabeling (fluorophore conjugated primary antibody) the last washing was immediately followed by the mounting. Otherwise, the sample was incubated with the secondary antibody diluted in the same way in 1% (w/v) BSA in PBS either at room temperature for 1h or at 4 °C overnight. The final concentration was as well 1-5 μg/ml. According to the incubation of the primary antibody this was followed by several washes 1% (w/v) BSA in PBS. An additional negative control where the cells were incubated with the secondary antibody only was always prepared when a primary antibody was used for the first time. If a secondary antibody was used for the first time a positive control with a commercial secondary antibody was prepared along with the regular sample.

5.5.3 Immunofluorescence

Mounting

Depending on the microscope and the dyes different mounting media were used. For wide field fluorescence and confocal microscopy the widely used Mowiol supplemented with DABCO (25% (w/v) glycerol, 9% (w/v) Mowiol 4-88, 0.1 M Tris/HCl pH 8.5, 0.1% (w/v) 1,4-diazabicyclo[2.2.2]octane) was sufficient. DABCO is an antioxidant quenching the triplet state of the dyes such increasing their lifetime. To mount the sample it was simply put onto a droplet of Mowiol dripped on a microscope slide. However, the refractive index of the unpolymerized Mowiol is about 1.47 (due to glycerol) which results in an refractive index mismatch between the mounted sample on the one hand and glass and the immersion oil ($n_d = 1, 51$) on the other hand leading to (???). To overcome this mismatch another mounting medium, TDE (2,2'thiodiethanol), was used. By adding different amounts of an aqueous solution like PBS the refractive index can be matched linearly ranging from 1.33 (water) over 1.47 (glycerol) to 1.51 (glass and immersion oil). For $n_d = 1, 515$ to 97% (v/v) TDE 2% (v/v) PBS were added. If triplet quencher were desired they had to be dissolved in the aqueous buffer first. Due to the high viscosity of TDE one cannot simply embed the sample as described above for Mowiol. Therefore various dilutions with increasing TDE content (10, 25, 50, 80 and 97%) were used to exchange the water in the cells with TDE in a stepwise manner. For TDE does not polymerize but stays liquid the sample needed to be sealed with nail polish. As already mentioned at the beginning of this section the choice of the mounting medium was also dependent on the dyes. Some dyes showed an enormous increase of fluorescence intensity in TDE whereas others showed a strong decrease.

In general, all samples were not stored for long time but measured (see 5.6) either the same day or the following days.

5.6 Light microscopy

5.6.1 Wide field Fluorescence microscopy

All wide field fluorescence and transmission light microscopy was performed with the upright Leica DM6000 (Leica Microsystems, Wetzlar, Germany). It is equipped with dry lenses (HC PL FLUOTAR 5x NA 0.15; HC PL FLUOTAR 10x NA 0.3; HC PL FLUOTAR 20x NA 0.5; HC PL FLUOTAR 40x NA 0.75), a water immersion lens (HCX PL APO 63x NA 1.2 W CORR) and an oil immersion lens (HCX PL APO 100x NA 1.4). As a light source for transmission light microscopy a halogen lamp was used whereas the illumination for fluorescence microscopy was carried out by a metal halide lamp (120 W). Different excitation filters were placed in the beam path before a dichroic mirror that guides the selected excitation wavelength through the lens onto the specimen. The emitted fluorescence passes the dichroic mirror and is detected by a CCD camera where a different set of emission filters was placed before.

5.6.2 Confocal microscopy

For the documentation of the fluorescently labeled samples, a confocal laser scanning microscope (CSLM) was used (TCS SP5, Leica Microsystems, Wetzlar, Germany). All recordings were done using a pinhole diameter of one Airy unit ($1.22x\lambda/NA$) and a scanning speed of 400 lines per second. The voxel size was typically set to 50 nm x 50 nm x 150 nm (X x Y x Z) according to the Nyquist Sampling Theorem. The used fluorophores were excited with the accordant laser wavelengths close to their absorption maximum. Four different laser sources with one or more distinct wavelengths are implemented in this setup: UV diode laser (405 nm); argon laser (458 nm, 476 nm, 488 nm, 496 nm and 514 nm); diode-pumped solid-state laser (DPSS, 561 nm); Helium-Neon laser (HeNe, 633 nm) The detection was carried out by using photomultiplier tubes (PMTs) which were operated in the dynamic range. The detection range was set around the emission maximum. Instead of using filters for separating the excitation from the emission spectrum an acousto optical tunable filter (AOTF) was used. To improve the signal-to-noise ratio (SNR) each line was scanned 4-times and averaged. The TCS SP5 is equipped with several different objective lenses: HC PL APO CS 10x NA 0.4 DRY; HCX PL APO CS 20x NA 0.7 multi immersion lens; HCX PL APO CS 40x NA 1.25 OIL; HCX PL APO CS

63x NA 1.4 OIL; HCX PL APO lambda blue 63x NA 1.4 OIL UV; HCX PL APO CS 63x NA 1.2 WATER. Most samples were imaged using the 63x NA 1.4 oil immersion lens.

5.6.3 STED-FCS setup

A standard epi-illuminated confocal microscope was modified for the STED-FCS experiments by Dr. Christian Ringemann (Ringemann, 2008). In short, fluorescence excitation was performed with a 633 nm pulsed laser diode ($\sim$ 80 ps pulse width, LDH-P-635, Picoquant, Berlin, Germany) or a continuous-wave (CW) laser diode (FiberTEC635, AMS Technologies, Munich, Germany). The STED light was supplied by a Titan:Sapphire laser system (MaiTai, Spectra-Physics, Mountain View, CA, USA; or Mira 900F, Coherent, Santa Barbara, USA) running at 750-780 nm and with a repetition rate of 76 and 80 MHz, respectively, or the Mira 900F system operated in a CW mode at 780 nm. The power of the STED laser light was controlled and stabilized by a laser power controller unit (LPC, Brockton Electronics, Brockton, MA, USA). The pulses were stretched to about 250-300 ps by coupling the STED laser into a 120 m long single-mode fiber (AMS Technologies, Munich, Germany). Both laser beams were spatially overlaid (dicroic filters, AHF Analysentechnik, Tübingen, Germany), directed on a beam scanning device (PSH 10/2, Piezosysteme Jena, Jena, Germany) and imaged into the microscope (DMIRBE, Leica Microsystems, Mannheim, Germany). An oil immersion objective (PL APON 60x NA 1.42, Olympus Japan, or HCX PL APO NA 1.4, Leica) was used to focus the laser light to a diffraction-limited spot on the sample and to collect the fluorescence emission. The doughnut-shaped focal spot with a central zero of the STED light was produced by introducing a phase plate into the beam path.

The fluorescence was imaged back over the beam scanning device and coupled into a multi-mode fiber splitter (Fiber Optic Network Technology, Surrey, Canada) with an aperture size corresponding to 0.8x the magnified Airy disc. The 50:50 split fluorescence signals were then detected by two avalanche photo diodes (APD, SPCM-AQR-13-FC, Perkin Elmer Optoelectronics, Fremont, CA). The fluorescence counts were further processed by a hardware correlator card (Flex02-01D, Correlator.com, NJ, USA) for FCS. For calibration of the STED microscope with lateral confinement the diameters of the effective fluorescence foci was determined by scanning $\sim$ 20 nm large fluorescent crimson beads (Invitrogen, Carlsbad, CA, USA).

Materials and Methods

The samples for the STED-FCS experiments were mounted in aqueous solutions. Usually HEPES buffered culture medium without serum and phenol red was used to allow for a prolonged cell viability. Fluorescence bursts, originating from single dye molecules diffusing through the detection volume, are detected. The burst length depends on the diffusion path through the focal spot. By decreasing the detection volume by a doughnut shaped STED pattern, also decreases the burst lengths, since the dye molecules pass this effective detection volume faster than the confocal volume. This decrease appears after FCS analysis in a decreased lateral diffusion time proportional to the STED intensity.

5.6.4 Supercontinuum STED

The setup of the supercontinuum STED microscope is described in (Wildanger et al., 2008) in detail. In brief, both the excitation and the STED pulse are provided by a supercontinuum (SC) fiber laser (SC-450 HP, Fianium, Southampton, UK). The pulse length of a 20 nm spectral band of the supercontinuum was measured to be (82 ± 10) ps and is therefore already appropriate for STED without the need of further stretching. Since the excitation and the STED pulses originate from the same resonator, they are inherently synchronized. Their simultaneous arrival at the focal plane can be adjusted by equalizing the optical path lengths with a comfortable tolerance of ~ 0.5 cm. Thus, apart from the laser source and the optical components typically required for a scanning (confocal) microscope, this STED microscope design necessitates only few additional components.

To take full advantage of the full spectral power density of the supercontinuum laser, besides the excitation wavelength, two STED beams are derived from the source. The wavelength needed for the organic dye Atto590 (Atto-Tec, Siegen, Germany) are (570 nm) for the excitation and (700 nm) for the STED beam, respectively. The laser output is first filtered to remove the infrared components of the spectrum and is then split using a polarizing beam splitter cube to provide two orthogonally polarized but otherwise equal beams. The excitation laser beam is derived from the s-polarized beam using a short pass dichroic beamsplitter (Z670SPRDC, AHF Analysentechnik GmbH, Tübingen, Germany) positioned prior to the monochromator. It is then directed through a bandpass filter and coupled into a polarization-preserving single mode fiber. At the fiber outputs, the two STED laser beams are separately collimated, recombined with a polarizing beam splitter cube and coupled into the objective lens with a dichroic beam splitter. Likewise, the

5.6.4 Supercontinuum STED

excitation light leaving the optical fiber is collimated and is coupled into the objective lens with a second dichroic mirror (Z568RDC, AHF Analysentechnik GmbH, Tübingen, Germany). Careful adjustment of the orientation and tilt of a superachromatic quarter wave plate (RSU1.4.15, B. Halle Nachfl.GmbH, Berlin, Germany) placed in front of the objective lens renders circular polarization for both STED beams. The beams were separately coupled into polarization-maintaining single mode optical fibers. The fiber outputs were collimated and colinearly coupled into an oil immersion objective lens (PL APO 100x/1.40-0.7 Oil, Leica Microsystems, Wetzlar, Germany) using two dichroic mirrors. To create a focal doughnut, the expanded STED beam first passed through an achromatic half-wave plate (500-900 nm, B. Halle Nachfl. GmbH, Berlin, Germany), and subsequently through a vortex phase plate (RPC Photonics, Rochester, NY, USA) imprinting a helical phase ramp of 2π onto the wavefront. An achromatic quarterwave plate (500-900 nm, B. Halle GmbH) placed close to the back aperture of the objective lens afforded circular polarization of both the excitation and the STED beams which causes all vectorial components of the STED beam to interfere destructively at the geometric focal point, thus creating a deep doughnut minimum. Also, photoselection is minimized by circular polarization.

The collected fluorescence was focused into a multimode optical fiber (62.5 μm / 0.27 NA, M31L01, Thorlabs) which acted as a confocal pinhole of 1.12-1.40 times the size of an Airy disc. The fluorescence photons were registered with an avalanche photodiode module (SPCMAQRH-13-FC, Perkin Elmer, Vaudreuil, Quebec, Canada) connected to a time-correlated single-photon counting board (SPC-830, Becker and Hickl GmbH, Berlin, Germany). The image acquisition was performed by scanning the sample with a 3D piezo stage (NanoBlock, Melles Griot GmbH, Bensheim, Germany). Typically, images were acquired with a 10-20 nm pixel size and pixel dwell times of 0.2-1.0 ms. The STED and confocal reference images were recorded sequentially.

Materials and Methods

6 Abbreviations

AMPK	Adenosine monophosphate activated protein kinase
β-CD	β-Cyclodextrin
hBDNF	human Brain derived neurotrophic factor
BSA	Bovine serum albumin
CDK5	Cyclin-dependent kinase 5
Cer	Ceramide = sphingosine + fatty acid
C, Chol	Cholesterol
COase	Cholesterol oxidase
DCX	Doublecortin
DMEM	Dulbecco's modified Eagle medium
DOPE	1,2-dioctadecenoyl (=dioleoyl)-phosphatidylethanolamine
FCS	Fluorescence Correlation Spectroscopy
FND	Fluorescent nanodiamond
GM1	mono-sialo ganglioside
HBP	Hexosamine biosynthetic pathway
HDMEM	HEPES buffered DMEM
IF	Intermediate filament
IgG	Immunoglobulin G
IgM	Immunoglobulin M
I-type	Morphologic intermediate cell type
JNK	c-Jun N-terminal kinase
KSP	Lysine-serine-proline
MAPK	Mitogen activated protein kinase
NF	Neurofilament
NFH	Neurofilament heavy subunit
NFL	Neurofilament light subunit
NFM	Neurofilament medium subunit
NHS	N-hydroxysuccinimide
N-type	Neuronal cell type
N-V	Nitrogen-vacancy
O-GlcNAc	O-linked-β-N-acetylglucosamine

Abbreviations

OGT	O-linked-β-N-acetylglucosamine transferase
PC	Phosphatidylcholine
PE	Phosphatidylethanolamine
PI	Phosphatidylinositol
PS	Phosphatidylserine
PSF	Point spread function
PTM	Post-translational modification
RA	(here: all-*trans*) Retinoic acid
RESOLFT	Reversible saturated optical fluorescence transitions
SAPK	Stress activated protein kinase
SDS-PAGE	Sodium dodecyl sulfate polyacrylamide gl electrophoresis
Sf	Serum free
SM	Sphingomyelin
SMCC	Succinimidyl-4-(N-maleimidomethyl)cyclohexane-1-carboxylate
SNR	Signal-to-noise ratio
STED	Stimulated emission depletion
S-type	Substrate adherent cell type
TMA-DPH	N,N,N-Trimethyl-4-(6-phenyl-1,3,5-hexatrien-1-yl)phenylammonium p-toluenesulfonate
TPA	12-tetradecanoyl-13-acetyl-β-phorbol
TrkA	Tyrosine receptor kinase A
TrkB	Tyrosine receptor kinase B

Bibliography

Abbe, E. 1873. Beiträge zur theorie des mikroskops und der mikroskopischen wahrnehmung. *Arch. Mikr. Anat.*, 9:413–468.

Abbud, W., Habinowski, S., Zhang, J.-Z., Kendrew, J., Elkairi, F. S., Kemp, B. E., Witters, L. A., and Ismail-Beigi, F. 2000. Stimulation of amp-activated protein kinase (ampk) is associated with enhancement of glut1-mediated glucose transport. *Archives of Biochemistry and Biophysics*, 380(2):347–352.

Ackerley, S., Grierson, A. J., Banner, S., Perkinton, M. S., Brownlees, J., Byers, H. L., Ward, M., Thornhill, P., Hussain, K., Waby, J. S., Anderton, B. H., Cooper, J. D., Dingwall, C., Leigh, P. N., Shaw, C. E., and Miller, C. C. J. 2004. p38 alpha stress-activated protein kinase phosphorylates neurofilaments and is associated with neurofilament pathology in amyotrophic lateral sclerosis. *Molecular and Cellular Neuroscience*, 26(2):354–364.

Ackerley, S., Thornhill, P., Grierson, A. J., Brownlees, J., Anderton, B. H., Leigh, P. N., Shaw, C. E., and Miller, C. C. J. 2003. Neurofilament heavy chain side arm phosphorylation regulates axonal transport of neurofilaments. *Journal of Cell Biology*, 161(3):489–495.

Adler, V., Schaffer, A., Kim, J., Dolan, L., and Ronai, Z. 1995. Uv irradiation and heat shock mediate jnk activation via alternate pathways. *J. Biol. Chem.*, 270(44):26071–26077.

Amos, W. B., White, J. G., and Fordham, M. 1987. Use of confocal imaging in the study of biological structures. *Appl. Optics*, 26:3239–3243.

Arcangeli, A., Rosati, B., Crociani, O., Cherubini, A., Fontana, L., Passani, B., Wanke, E., and Olivotto, M. 1999. *Journal of Neurobiology*, 40(2):214–225.

Bibliography

Archer, D. R., Watson, D. F., and Griffin, J. W. 1994. Phosphorylation-dependent immunoreactivity of neurofilaments and the rate of slow axonal-transport in the central and peripheral axons of the rat dorsal-root ganglion. *Journal of Neurochemistry*, 62(3):1119–1125.

Avery, J., Ellis, D. J., Lang, T., Holroyd, P., Riedel, D., Henderson, R. M., Edwardson, J. M., and Jahn, R. 2000. A Cell-free System for Regulated Exocytosis in PC12 Cells. *J. Cell Biol.*, 148(2):317–324.

Axelrod, D., Koppel, D. E., Schlessinger, J., Elson, E., and Webb, W. W. 1976. Mobility measurement by analysis of fluorescence photobleaching recovery kinetics. *Biophysical Journal*, 16(9):1055–1069.

Ayuyan, A. and Cohen, F. 2006. Lipid peroxides promote large rafts: Effects of excitation of probes in fluorescence microscopy and electrochemical reactions during vesicle formation. *Biophys. J.*, 91:2172–2183.

Bacia, K., Scherfeld, D., Kahya, N., and Schwille, P. 2004. Fluorescence correlation spectroscopy relates rafts in model and native membranes. *Biophys. J.*, 87:1034–1043.

Baquer, N. Z., Taha, A., Kumar, P., McLean, P., Cowsik, S. M., Kale, R. K., Singh, R., and Sharma, D. 2009. A metabolic and functional overview of brain aging linked to neurological disorders. *Biogerontology*, 10(4):377–413.

Betzig, E., Patterson, G., Sougrat, R., Lindwasser, O., Olenych, S., Bonifacino, J., Davidson, M., Lippincott-Schwartz, J., and Hess, H. 2006. Imaging intracellular fluorescent proteins at nanometer resolution. *Science*, 313(5793):1642–1645.

Biedler, J. L., Helson, L., and Spengler, B. A. 1973. Morphology and growth, tumorigenicity, and cytogenetics of human neuroblastoma cells in continuous culture. *Cancer Research*, 33(11):2643–2652.

Birnboim, H. and Doly, J. 1979. A rapid alkaline extraction procedure for screening recombinant plasmid DNA. *Nucl. Acids Res.*, 7(6):1513–1523.

Bradford, M. M. 1976. A rapid and sensitive method for the quantitation of microgram quantities of protein utilizing the principle of protein-dye binding. *Anal Biochem.*, pages 248–54.

Breaker, R. R. 2004. Natural and engineered nucleic acids as tools to explore biology. 432(7019):838–845.

Bretschneider, S., Eggeling, C., and Hell, S. W. 2007. Breaking the diffraction barrier in fluorescence microscopy by optical shelving. *Phys. Rev. Lett.*, 98:218103.

Brownlees, J., Yates, A., Bajaj, N. P., Davis, D., Anderton, B. H., Leigh, P. N., Shaw, C. E., and Miller, C. C. J. 2000. Phosphorylation of neurofilament heavy chain sidearms by stress activated protein kinase-1b/jun n-terminal kinase-3. *Journal of Cell Science*, 113(3):401–407.

Bu, X., Huang, P., Qi, Z., Zhang, N., Han, S., Fang, L., and Li, J. 2007. Cell type-specific activation of p38 mapk in the brain regions of hypoxic preconditioned mice. *Neurochemistry International*, 51(8):459–466.

Buki, A., Okonkwo, D. O., Wang, K. K. W., and Povlishock, J. T. 2000. Cytochrome c release and caspase activation in traumatic axonal injury 20024057. *J. Neurosci.*, 20(8):2825–2834.

Caroni, P. and Golub, T. 2002. Actin cytoskeleton regulation at pi(4,5)p2 rafts. *Molecular Biology of the Cell*, 13:793.

Carpentier, A., Balitrand, N., RochetteEgly, C., Shroot, B., Degos, L., and Chomienne, C. 1997. Distinct sensitivity of neuroblastoma cells for retinoid receptor agonists: evidence for functional receptor heterodimers. *Oncogene*, 15(15):1805–1813.

Castagne, D., Fillet, M., Delattre, L., Evrard, B., Nusgens, B., and Piel, G. 2008. Study of the cholesterol extraction capacity of beta-cyclodextrin and its derivatives, relationships with their effects on endothelial cell viability and on membrane models. *J Incl phenom Macrocycl Chem*, 63:225–231.

Castillo, M. R. and Babson, J. R. 1998. Ca2+-dependent mechanisms of cell injury in cultured cortical neurons. *Neuroscience*, 86(4):1133–1144.

Chao, J. I., Perevedentseva, E., Chung, P. H., Liu, K. K., Cheng, C. Y., Chang, C. C., and Cheng, C. L. 2007. Nanometer-sized diamond particle as a probe for biolabeling. *Biophysical Journal*, 93(6):2199–2208.

Bibliography

Cheung, W. D. and Hart, G. W. 2008. Amp-activated protein kinase and p38 mapk activate o-glcnacylation of neuronal proteins during glucose deprivation. *J. Biol. Chem.*, 283(19):13009–13020.

Cochard, P. and Paulin, D. 1984. Initial expression of neurofilaments and vimentin in the central and peripheral nervous-system of the mouse embryo invivo. *Journal of Neuroscience*, 4(8):2080–2094.

Coons, A. H. and Kaplan, M. H. 1950. Localization of antigen in tissue cells .2. improvements in a method for the detection of antigen by means of fluorescent antibody. *Journal of Experimental Medicine*, 91(1):1–13.

Culmsee, C., Monnig, J., Kemp, B. E., and Mattson, M. P. 2001. Amp-activated protein kinase is highly expressed in neurons in the developing rat brain and promotes neuronal survival following glucose deprivation. *Journal of Molecular Neuroscience*, 17(1):45–58.

Dagon, Y., Avraham, Y., Magen, I., Gertler, A., Ben-Hur, T., and Berry, E. M. 2005. Nutritional status, cognition, and survival - a new role for leptin and amp kinase. *Journal of Biological Chemistry*, 280(51):42142–42148.

Davies, G. and Hamer, M. F. 1976. Optical studies of 1.945 ev vibronic band in diamond. *Proceedings of the Royal Society of London Series a-Mathematical Physical and Engineering Sciences*, 348(1653):285–298.

de Almeida, R. F., Loura, L., Fedorov, A., and Prieto, M. 2005. Lipid rafts have different sizes depending on membrane composition: A time-resolved fluorescence resonance energy transfer study. *J. Mol. Biol.*, 346:1109–1120.

Deng, Y. Q., Li, B., Liu, F., Iqbal, K., Grundke-Iqbal, I., Brandt, R., and Gong, C. X. 2008. Regulation between o-glcnacylation and phosphorylation of neurofilament-m and their dysregulation in alzheimer disease. *Faseb Journal*, 22(1):138–145.

Denk, W., Strickler, J. H., and Webb, W. W. 1990. Two-photon laser scanning fluorescence microscopy. *Science*, 248:73–76.

Devaux, P. F. 1993. Lipid transmembrane asymmetry and flip-flop in biological-membranes and in lipid bilayers. *Current Opinion in Structural Biology*, 3(4):489–494.

Diakowski, W., Ozimek, L., Bielska, E., Bem, S., Langner, M., and Sikorski, A. F. 2006. Cholesterol affects spectrin-phospholipid interactions in a manner different from changes resulting from alterations in membrane fluidity due to fatty acyl chain composition. *Biochimica et Biophysica Acta (BBA) - Biomembranes*, 1758(1):4–12.

DiazNido, J., Ulloa, L., Sanchez, C., and Avila, J. 1996. The role of the cytoskeleton in the morphological changes occurring during neuronal differentiation. *Seminars in Cell & Developmental Biology*, 7(5):733–739.

Dietrich, C., Bagatolli, L. A., Volovyk, N. L., Thompson, N. L., Levi, M., Jacobson, K., and Gratton, E. 2001. Lipid rafts reconstituted in model membranes. *Biophys J*, 80:1417–1428.

Dong, D. L. Y., Xu, Z. S., Chevrier, M. R., Cotter, R. J., Cleveland, D. W., and Hart, G. W. 1993. Glycosylation of mammalian neurofilaments - localization of multiple o-linked n-acetylglucosamine moieties on neurofilament polypeptides-l and polypeptides-m. *Journal of Biological Chemistry*, 268(22):16679–16687.

Durig, U., Pohl, D. W., and Rohner, F. 1986. Near-field optical-scanning microscopy. *Journal of Applied Physics*, 59(10):3318–3327.

Edidin, M. 2001. Shrinking patches and slippery rafts: scales of domains in the plasma membrane. *Trends in Cell Biology*, 11(12):492–496.

Edidin, M. 2003. The state of lipid rafts: from mode membranes to cells. *Annu. Rev. Biophys. Biomol. Struct*, 32:257–283.

Edidin, M. and Stroynowski, I. 1991. Differences between the lateral organization of conventional and inositol phospholipid-anchored membrane-proteins - a further definition of micrometer scale membrane domains. *Journal of Cell Biology*, 112(6):1143–1150.

Eggeling, C., Berger, S., Brand, L. ., Fries, J., Schaffer, J., Volkmer, A., and Seidel, C. A. M. 2001. Data registration and selective single-molecule analysis using multi-parameter fluorescence detection. *J. Biotechnol.*, 86:163–180.

Eggeling, C., Ringemann, C., Medda, R., Schwarzmann, G., Sandhoff, K., Polyakova, S., Belov, V., Hein, B., Middendorff, C. v., Schönle, A., and Hell, S. W. 2009. Direct

Bibliography

observation of the nanoscale dynamics of membrane lipids in a living cell. *Nature*, 457:1159–1163.

Eggeling, C., Widengren, J., Rigler, R., and Seidel, C. A. M. 1998. Photobleaching of fluorescent dyes under conditions used for single-molecule detection: Evidence of two-step photolysis. *Anal. Chem.*, 70:2651–2659.

Ehrhard, P. B., Ganter, U., Schmutz, B., Bauer, J., and Otten, U. 1993. Expression of low-affinity ngf receptor and trkb mrna in human sh-sy5y neuroblastoma cells. *FEBS Letters*, 330(3):287 – 292.

Elhanany, E., Jaffe, H., Link, W. T., Sheeley, D. M., Gainer, H., and Pant, H. C. 1994. Identification of endogenously phosphorylated ksp sites in the high-molecular-weight rat neurofilament protein. *Journal of Neurochemistry*, 63(6):2324–2335.

Elson, E. and Magde, D. 1974. Fluorescence correlation spectroscopy i. conceptual basis an theory. *Biopolymers*, 13:1–27.

Encinas, M., Iglesias, M., Liu, Y., Wang, H., Muhaisen, A., Cena, V., Gallego, C., and Comella, J. X. 2000. Sequential treatment of sh-sy5y cells with retinoic acid and brain-derived neurotrophic factor gives rise to fully differentiated, neurotrophic factor-dependent, human neuron-like cells. *Journal of Neurochemistry*, 75(3):991–1003.

Faklaris, O., Garrot, D., Joshi, V., Druon, F., Boudou, J.-P., Sauvage, T., Georges, P., Curmi, P. A., and Treussart, F. 2008. Detection of single photoluminescent diamond nanoparticles in cells and study of the internalization pathway. *Small*, 4(12):2236–2239.

Farber, J. L. 1990. The role of calcium-ions in toxic cell injury. *Environmental Health Perspectives*, 84:107–111.

Favre, B., Zolnierowicz, S., Turowski, P., and Hemmings, B. A. 1994. The catalytic subunit of protein phosphatase 2a is carboxyl-methylated in-vivo. *Journal of Biological Chemistry*, 269(23):16311–16317.

Feder, T. J., Brust-Mascher, I., Slattery, J. P., Baird, B. A., and Webb, W. W. 1996. Constrained diffusion or immobile fraction on cell surfaces: A new interpretation. *Biophys. J.*, 70:2767–2773.

Felton, S., Edmonds, A. M., Newton, M. E., Martineau, P. M., Fisher, D., and Twitchen, D. J. 2008. Electron paramagnetic resonance studies of the neutral nitrogen vacancy in diamond. *Physical Review B (Condensed Matter and Materials Physics)*, 77(8):081201–4.

Fishman, P. H. 1982. Role of membrane gangliosides in the binding and action of bacterial toxins. *Journal of Membrane Biology*, 69(2):85–97.

Fladeby, C., Skar, R., and Serck-Hanssen, G. 2003. Distinct regulation of glucose transport and glut1/glut3 transporters by glucose deprivation and igf-i in chromaffin cells. *Biochimica et Biophysica Acta (BBA) - Molecular Cell Research*, 1593(2-3):201–208.

Fliegner, K. H. and Liem, R. K. H. 1991. Cellular and molecular-biology of neuronal intermediate filaments. *International Review of Cytology-a Survey of Cell Biology*, 131:109–167.

Forno, L. S., Sternberger, L. A., Sternberger, N. H., Strefling, A. M., Swanson, K., and Eng, L. F. 1986. Reaction of lewy bodies with antibodies to phosphorylated and non-phosphorylated neurofilaments. *Neuroscience Letters*, 64(3):253–258.

Fujiwara, T., Ritchie, K., Murakoshi, H., Jacobson, K., and Kusumi, A. 2002. Phospholipids undergo hop diffusion in compartmentalized cell membrane. *J. Cell Biol.*, 157(6):1071–1081.

Gadalla, A. E., Pearson, T., Currie, A. J., Dale, N., Hawley, S. A., Sheehan, M., Hirst, W., Michel, A. D., Randall, A., Hardie, D. G., and Frenguelli, B. G. 2004. Aica riboside both activates amp-activated protein kinase and competes with adenosine for the nucleoside transporter in the ca1 region of the rat hippocampus. *Journal of Neurochemistry*, 88(5):1272–1282.

Geisler, N. and Weber, K. 1981. Self-assembly invitro of the 68,000 molecular-weight component of the mammalian neurofilament triplet proteins into intermediate-sized filaments. *Journal of Molecular Biology*, 151(3):565–571.

Giasson, B. I. and Mushynski, W. E. 1996. Aberrant stress-induced phosphorylation of perikaryal neurofilaments. *Journal of Biological Chemistry*, 271(48):30404–30409.

Bibliography

Girotti, A. W. 2001. Photosensitized oxidation of membrane lipids: reaction pathways, cytotoxic effects, and cytoprotective mechanisms. *Journal of Photochemistry and Photobiology B: Biology*, 63(1-3):103–113.

Glick, R. D., Medary, I., Aronson, D. C., Scotto, K. W., Swendeman, S. L., and La Quaglia, M. P. 2000. The effects of serum depletion and dexamethasone on growth and differentiation of human neuroblastoma cell lines. *Journal of Pediatric Surgery*, 35(3):465–472.

Glicksman, M. A., Soppet, D., and Willard, M. B. 1987. Posttranslational modification of neurofilament polypeptides in rabbit retina. *Journal of Neurobiology*, 18(2):167–196.

Gong, C. X., Shaikh, S., Wang, J. Z., Zaidi, T., Grundkeiqbal, I., and Iqbal, K. 1995. Phosphatase-activity toward abnormally phosphorylated-tau - decrease in alzheimer-disease brain. *Journal of Neurochemistry*, 65(2):732–738.

Gong, C. X., Singh, T. J., Grundkeiqbal, I., and Iqbal, K. 1993. Phosphoprotein phosphatase-activities in alzheimer-disease brain. *Journal of Neurochemistry*, 61(3):921–927.

Grant, P. and Pant, H. C. 2000. Neurofilament protein synthesis and phosphorylation. *Journal of Neurocytology*, 29(11-12):843–872.

Gruber, A., Drabenstedt, A., Tietz, C., Fleury, L., Wrachtrup, J., and Borczyskowski, C. v. 1997. Scanning confocal optical microscopy and magnetic resonance on single defect centers. *Science*, 276(5321):2012–2014.

Guan, R. J., Khatra, B. S., and Cohlberg, J. A. 1991. Phosphorylation of bovine neurofilament proteins by protein-kinase fa (glycogen-synthase kinase-3). *Journal of Biological Chemistry*, 266(13):8262–8267.

Gudas, L. J. 1994. Retinoids and vertebrate development. *Journal of Biological Chemistry*, 269(22):15399–15402.

Guidato, S., Tsai, L. H., Woodgett, J., and Miller, C. C. J. 1996. Differential cellular phosphorylation of neurofilament heavy side-arms by glycogen synthase kinase-3 and cyclin-dependent kinase-5. *Journal of Neurochemistry*, 66(4):1698–1706.

Guo, H. and Damuni, Z. 1993. Autophosphorylation-activated protein-kinase phosphorylates and inactivates protein phosphatase-2a. *Proceedings of the National Academy of Sciences of the United States of America*, 90(6):2500–2504.

Haines, T. H. 2001. Do sterols reduce proton and sodium leaks through lipid bilayers? *Progress in Lipid Research*, 40(4):299–324.

Hankins, J. 2006. The role of albumin in fluid and electrolyte balance. *J Infus Nurs*, 29(5).

Hannun, Y. A. and Obeid, L. M. 2002. The ceramide-centric universe of lipid-mediated cell regulation: Stress encounters of the lipid kind. *J. Biol. Chem.*, 277(29):25847–25850.

Hardie, D. G. and Hawley, S. A. 2001. Amp-activated protein kinase: the energy charge hypothesis revisited. *Bioessays*, 23(12):1112–1119.

Heerklotz, H. 2002. Triton promotes domain form in lipid raft mixtures. *Biophys J*, 83:2693–2701.

Heidemann, S. R. 1996. Cytoplasmic mechanisms of axonal and dendritic growth in neurons. In *International Review of Cytology - a Survey of Cell Biology, Vol 165*. ACADEMIC PRESS INC.

Hell, S. W. 2007. Far-field optical nanoscopy. *Science*, 316(5828):1153–1158.

Hell, S. W., Dyba, M., and Jakobs, S. 2004. Concepts for nanoscale resolution in fluorescence microscopy. *Curr. Opin. Neurobio.*, 14(5):599–609.

Hell, S. W. and Kroug, M. 1995. Ground-state depletion fluorescence microscopy, a concept for breaking the diffraction resolution limit. *Appl. Phys. B*, 60:495–497.

Hell, S. W. and Wichmann, J. 1994. Breaking the diffraction resolution limit by stimulated emission: stimulated emission depletion fluorescence microscopy. *Opt. Lett.*, 19(11):780–782.

Hess, S. T., Girirajan, T. P. K., and Mason, M. D. 2006. Ultra-high resolution imaging by fluorescence photoactivation localization microscopy. *Biophys. J.*, 91(11):4258–4272.

Bibliography

Hill, W. D., Arai, M., Cohen, J. A., and Trojanowski, J. Q. 1993. Neurofilament messenger-rna is reduced in parkinsons-disease substantia-nigra pars-compacta neurons. *Journal of Comparative Neurology*, 329(3):328–336.

Hisanaga, S., Gonda, Y., Inagaki, M., Ikai, A., and Hirokawa, N. 1990a. Effects of phosphorylation of the neurofilament l-protein on filamentous structures. *Cell Regulation*, 1(2):237–248.

Hisanaga, S., Ikai, A., and Hirokawa, N. 1990b. Molecular architecture of the neurofilament .1. subunit arrangement of neurofilament-l protein in the intermediate-sized filament. *Journal of Molecular Biology*, 211(4):857–869.

Hooke, R. 1665. *Micrographia: or, Some physiological descriptions of minute bodies made by magnifying glasses.* J. Martyn and J. Allestry, London, first edition.

Hooper, C., Killick, R., and Lovestone, S. 2008. The gsk3 hypothesis of alzheimer's disease. *Journal of Neurochemistry*, 104(6):1433–1439.

Hori, Y., Spurr-Michaud, S. J., Russo, C. L., Argueso, P., and Gipson, I. K. 2005. Effect of retinoic acid on gene expression in human conjunctival epithelium: Secretory phospholipase a2 mediates retinoic acid induction of muc16. *Invest. Ophthalmol. Vis. Sci.*, 46(11):4050–4061.

Hsueh, Y. W., Gilbert, K., Trandum, C., Zuckermann, M., and Thewalt, J. 2005. The effect of ergosterol on dipalmitoylphosphatidylcholine bilayers: a deuterium nmr and calorimetric study. *Biophys J*, 88:1799–1808.

Iakoubovskii, K., Adriaenssens, G. J., and Nesladek, M. 2000. Photochromism of vacancy-related centres in diamond. *Journal of Physics: Condensed Matter*, (2):189.

Iqbal, K. and Grundke-Iqbal, I. 2005. Metabolic/signal transduction hypothesis of alzheime's disease and other tauopathies. *Acta Neuropathologica*, 109(1):25–31.

Ishihara, H., Martin, B. L., Brautigan, D. L., Karaki, H., Ozaki, H., Kato, Y., Fusetani, N., Watabe, S., Hashimoto, K., Uemura, D., and Hartshorne, D. J. 1989. Calyculin-a and okadaic acid - inhibitors of protein phosphatase-activity. *Biochemical and Biophysical Research Communications*, 159(3):871–877.

Ishii, T., Haga, S., and Tokutake, S. 1979. Presence of neurofilament protein in alzheimer neurofibrillary tangles (ant) - immunofluorescent study. *Acta Neuropathologica*, 48(2):105–112.

Jacobson, K., Mouritsen, O., and Anderson, G. 2007. Lipid rafts: at a crossroad between cell biology and physics. *Nature Cell Biol.*, 9(1):7–14.

Jaffe, H., Veeranna, Shetty, K. T., and Pant, H. C. 1998. Characterization of the phosphorylation sites of human high molecular weight neurofilament protein by electrospray ionization tandem mass spectrometry and database searching. *Biochemistry*, 37(11):3931–3940.

James, W. 2001. Nucleic acid and polypeptide aptamers: a powerful approach to ligand discovery. *Current Opinion in Pharmacology*, 1(5):540–546.

Julien, J. P. and Mushynski, W. E. 1982. Multiple phosphorylation sites in mammalian neurofilament polypeptides. *Journal of Biological Chemistry*, 257(17):467–470.

Julien, J. P. and Mushynski, W. E. 1983. The distribution of phosphorylation sites among identified proteolytic fragments of mammalian neurofilaments. *Journal of Biological Chemistry*, 258(6):4019–4025.

Jung, C., Yabe, J. T., Lee, S., and Shea, T. B. 2000. Hypophosphorylated neurofilament subunits undergo axonal transport more rapidly than move extensively phosphorylated subunits in situ. *Cell Motility and the Cytoskeleton*, 47(2):120–129.

Kaplan, D. R., Matsumoto, K., Lucarelli, E., and Thielet, C. J. 1993. Induction of trkb by retinoic acid mediates biologic responsiveness to bdnf and differentiation of human neuroblastoma cells. *Neuron*, 11(2):321–331.

Kato, S. and Hirano, A. 1991. Ultrastructural identification of alzheimers neurofibrillary tangles (nft) in spinal-cords in guamanian patients with amyotrophic-lateral-sclerosis (als) and parkinsonism-dementia complex (pdc). *Journal of Neuropathology and Experimental Neurology*, 50(3):306–306.

Keller, J., Schönle, A., and Hell, S. W. 2007. Efficient fluorescence inhibition patterns for resolft microscopy. *Opt. Express*, 15(6):3361–3371.

Bibliography

Klar, T. and Hell, S. 1999. Subdiffraction resolution in far-field fluorescence microscopy. *Opt. Lett.*, 24(14):954–956.

Knoll, M. and Ruska, E. 1932. Das elektronenmikroskop. *Zeitschrift für Physik*, 78:318–339.

Kreppel, L. K. and Hart, G. W. 1999. Regulation of a cytosolic and nuclear o-glcnac transferase - role of the tetratricopeptide repeats. *Journal of Biological Chemistry*, 274(45):32015–32022.

Kurth-Kraczek, E. J., Hirshman, M. F., Goodyear, L. J., and Winder, W. W. 1999. 5' amp-activated protein kinase activation causes glut4 translocation in skeletal muscle. *Diabetes*, 48(8):1667–1671.

Kwik, J., Boyle, S., Fooksman, D., Margolis, L., Sheetz, M., and Edidin, M. 2003. Membrane cholesterol, lateral mobility, and the phosphatidylinositol 4,5-bisphosphatedependent organization of cell actin. *Proc. Natl. Acad. Sci. USA*, 100(24):13964–13969.

Laemmli, U. K. 1970. Cleavage of structural proteins during the assembly of the head of bacteriophage t4. *Nature*, pages 680–5.

Lakowicz, J. R. 1999. *Principles of fluorescence spectroscopy*. Kluwer Academic/Plenum, New York, 2nd edition.

Lee, M. K. and Cleveland, D. W. 1996. Neuronal intermediate filaments. *Annual Review of Neuroscience*, 19(1):187–217.

Levene, M., Korlach, J., Turner, S., Foquet, M., Craighead, H., and Webb, W. 2003. Zero-mode waveguides for single-molecule analysis at high concentrations. *Science*, 299:682–686.

Lewis, S. E. and Nixon, R. A. 1988. Multiple phosphorylated variants of the high molecular mass subunit of neurofilaments in axons of retinal cell neurons - characterization and evidence for their differential association with stationary and moving neurofilaments. *Journal of Cell Biology*, 107(6):2689–2701.

Li, B. S., Veeranna, Grant, P., and Pant, H. C. 1999a. Activation of mitogen-activated protein kinase (erk1 and erk2) cascade results in phosphorylation of nf-m tail domain in transfected nih 3t3 cells. *Journal of Neurochemistry*, 72:S59–S59.

Li, B. S., Veeranna, Grant, P., and Pant, H. C. 1999b. Calcium influx and membrane depolarization induce phosphorylation of neurofilament (nf-m) ksp repeats in pc12 cells. *Molecular Brain Research*, 70(1):84–91.

Li, B. S., Zhang, L., Gu, J. G., Amin, N. D., and Pant, H. C. 2000. Integrin alpha(1)beta(1)-mediated activation of cyclin-dependent kinase 5 activity is involved in neurite outgrowth and human neurofilament protein h lys-ser-pro tail domain phosphorylation. *Journal of Neuroscience*, 20(16):6055–6062.

Li, M., Guo, H., and Damuni, Z. 1995. Purification and characterization of 2 potent heat-stable protein inhibitors of protein phosphatase 2a from bovine kidney. *Biochemistry*, 34(6):1988–1996.

Li, S. P., Deng, Y. Q., Wang, X. C., Wang, Y. P., and Wang, J. Z. 2004. Melatonin protects sh-sy5y neuroblastoma cells from calyculin a-induced neurofilament impairment and neurotoxicity. *Journal of Pineal Research*, 36(3):186–191.

Lillemeier, B. F., Pfeiffer, J. R., Surviladze, Z., Wilson, B. S., and Davis, M. M. 2006. Plasma membrane-associated proteins are clustered into islands attached to the cytoskeleton. *Proceedings of the National Academy of Sciences of the United States of America*, 103(50):18992–18997.

Liu, C., Russell, R. M., and Wang, X. D. 2004. Low dose beta-carotene supplementation of ferrets attenuates smoke-induced lung phosphorylation of jnk, p38 mapk, and p53 proteins. *Journal of Nutrition*, 134(10):2705–2710.

Liu, W.-H., Horng, W.-C., and Tsai, M.-S. 1996. Bioconversion of cholesterol to cholest-4-en-3-one in aqueous/organic solvent two-phase reactors. *Enzyme and Microbial Technology*, 18(3):184–189.

Loubser, J. H. N. and Wyk, J. A. v. 1978. Electron spin resonance in the study of diamond. *Reports on Progress in Physics*, (8):1201.

Bibliography

Lovat, P. E., Pearson, A. D. J., Malcolm, A., and Redfern, C. P. F. 1993. Retinoic acid receptor expression during the in-vitro differentiation of human neuroblastoma. *Neuroscience Letters*, 162(1-2):109–113.

Love, D. C. and Hanover, J. A. 2005. The hexosamine signaling pathway: deciphering the "o-glcnac code". *Sci STKE*, 2005(312):re13.

Ludemann, N., Clement, A., Hans, V. H., Leschik, J., Behl, C., and Brandt, R. 2005. O-glycosylation of the tail domain of neurofilament protein m in human neurons and in spinal cord tissue of a rat model of amyotrophic lateral sclerosis (als). *Journal of Biological Chemistry*, 280(36):31648–31658.

Madge, D. 1976. Chemical kinetics and fluorescence correlation spectroscopy. *Quart. Rev. Biophys.*, 9(1):35–47.

Madge, D., Elson, E., and Webb, W. 1974. Fluorescence correlation spectroscopy ii. an experimental realization. *Biopolymers*, 13:29–61.

Manetto, V., Sternberger, N. H., Perry, G., Sternberger, L. A., and Gambetti, P. 1988. Phosphorylation of neurofilaments is altered in amyotrophic lateral sclerosis. *Journal of Neuropathology and Experimental Neurology*, 47(6):642–653.

Marsin, A.-S., Bertrand, L., Rider, M., Deprez, J., Beauloye, C., Vincent, M., Van den Berghe, G., Carling, D., and Hue, L. 2000. Phosphorylation and activation of heart pfk-2 by ampk has a role in the stimulation of glycolysis during ischaemia. *Current Biology*, 10(20):1247–1255.

McCullough, L. D., Zeng, Z. Y., Li, H., Landree, L. E., McFadden, J., and Ronnett, G. V. 2005. Pharmacological inhibition of amp-activated protein kinase provides neuroprotection in stroke. *Journal of Biological Chemistry*, 280(21):20493–20502.

Medda, R., Jakobs, S., Hell, S. W., and Bewersdorf, J. 2006. 4pi microscopy of quantum dot-labeled cellular structures. *J. Struct. Biol.*, 156(3):517–523.

Meyer, L., Wildanger, D., Medda, R., Punge, A., Rizzoli, S. O., Donnert, G., and Hell, S. W. 2008. Dual-color sted microscopy at 30 nm focal plane resolution. (submitted).

Michie, K. A. and Löwe, J. 2006. Dynamic filaments of the bacterial cytoskeleton. *Annual Review of Biochemistry*, 75(1):467 LP – 492.

Minsky, M. 1961. Microscopy apparatus, us patent 3,013,467.

Mita, Y. 1996. Change of absorption spectra in type-ib diamond with heavy neutron irradiation. *Physical Review B*, 53:11360 – 11364.

Moerner, W. and Fromm, D. 2003. Methods of single-molecule fluorescence spectroscopy and microscopy. *Rev. Sci. Instrum.*, 74(8):3597–3619.

Mukai, H., Toshimori, M., Shibata, H., Kitagawa, M., Shimakawa, M., Miyahara, M., Sunakawa, H., and Ono, Y. 1996. Pkn associates and phosphorylates the head-rod domain of neurofilament protein. *Journal of Biological Chemistry*, 271(16):9816–9822.

Mukherjee, S. and Maxfield, F. R. 2004. Membrane domains. *Annual Review of Cell and Developmental Biology*, 20:839–866.

Murphy, R., Slayter, H., Schurtenberger, P., Chamberlin, R., Colton, C., and Yarmush, M. 1988. Size and structure of antigen-antibody complexes. electron microscopy and light scattering studies. *Biophysical Journal*, 54(1):45–56.

Nicotera, P., Bellomo, G., and Orrenius, S. 1990. The role of ca-2+ in cell killing. *Chemical Research in Toxicology*, 3(6):484–494.

Nishida, Y., Adati, N., Ozawa, R., Maeda, A., Sakaki, Y., and Takeda, T. 2008. Identification and classification of genes regulated by phosphatidylinositol 3-kinase- and trkb-mediated signalling pathways during neuronal differentiation in two subtypes of the human neuroblastoma cell line sh-sy5y. *BMC Research Notes*, 1(1):95.

Nitsche, A., Kurth, A., Dunkhorst, A., Panke, O., Sielaff, H., Junge, W., Muth, D., Scheller, F., Stocklein, W., Dahmen, C., Pauli, G., and Kage, A. 2007. One-step selection of vaccinia virus-binding dna aptamers by monolex. *Bmc Biotechnology*, 7.

Nixon, R. A., Brown, B. A., and Marotta, C. A. 1982. Posttranslational modification of a neurofilament protein during axoplasmic-transport - implications for regional specialization of cns axons. *Journal of Cell Biology*, 94(1):150–158.

Bibliography

Nixon, R. A., Lewis, S. E., Dahl, D., Marotta, C. A., and Drager, U. C. 1989. Early post-translational modifications of the three neurofilament subunits in mouse retinal ganglion cells: neuronal sites and time course in relation to subunit polymerization and axonal transport. *Brain Res Mol Brain Res*, 5(2):93–108.

Nixon, R. A. and Logvinenko, K. B. 1986. Multiple fates of newly synthesized neurofilament proteins - evidence for a stationary neurofilament network distributed nonuniformly along axons of retinal ganglion-cell neurons. *Journal of Cell Biology*, 102(2):647–659.

Nixon, R. A., Paskevich, P. A., Sihag, R. K., and Thayer, C. Y. 1994. Phosphorylation on carboxyl-terminus domains of neurofilament proteins in retinal ganglion-cell neurons in-vivo - influences on regional neurofilament accumulation, interneurofilament spacing, and axon caliber. *Journal of Cell Biology*, 126(4):1031–1046.

Nixon, R. A. and Shea, T. B. 1992. Dynamics of neuronal intermediate filaments - a developmental perspective. *Cell Motility and the Cytoskeleton*, 22(2):81–91.

O'Ferrall, E. K., Robertson, J., and Mushynski, W. E. 2000. Inhibition of aberrant and constitutive phosphorylation of the high-molecular-mass neurofilament subunit by cep-1347 (kt7515), an inhibitor of the stress-activated protein kinase signaling pathway. *Journal of Neurochemistry*, 75(6):2358–2367.

Orr, G., Hu, D., Özcelik, S., Opresko, L. K., Steven, W. H., and Colson, S. D. 2005. Cholesterol dictates the freedom of egf receptors and her2 in the plane of the membrane. *Biophysical Journal*, 89:1362–1373.

Orrit, M. and Bernard, J. 1990. Single pentacene molecules detected by fluorescence excitation in a p-terphenyl crystal. *Phys. Rev. Lett.*, 65:2716–2719.

Pachter, J. S. and Liem, R. K. H. 1985. Alpha-internexin, a 66-kd intermediate filament binding-protein from mammalian central nervous tissues. *Journal of Cell Biology*, 101(4):1316–1322.

Pagano, R. E., Sepanski, M. A., and Martin, O. C. 1989. Molecular trapping of a fluorescent ceramide analogue at the golgi apparatus of fixed cells: interaction with endogenous lipids provides a trans-golgi marker for both light and electron microscopy 10.1083/jcb.109.5.2067. *J. Cell Biol.*, 109(5):2067–2079.

Pahlman, S., Ruusala, A. I., Abrahamsson, L., Mattsson, M. E. K., and Esscher, T. 1984. Retinoic acid-induced differentiation of cultured human neuro-blastoma cells - a comparison with phorbolester-induced differentiation. *Cell Differentiation*, 14(2):135–144.

Palozza, P., Serini, S., and Calviello, G. 2006. Carotenoids as modulators of intracellular signaling pathways. *Current Signal Transduction Therapy*, 1(3):325–335.

Pan, J., Kao, Y.-L., Joshi, S., Jeetendran, S., DiPette, D., and Singh, U. S. 2005. Activation of rac1 by phosphatidylinositol 3-kinase <i>in vivo</i>: role in activation of mitogen-activated protein kinase (mapk) pathways and retinoic acid-induced neuronal differentiation of sh-sy5y cells. *Journal of Neurochemistry*, 93(3):571–583.

Pant, H. C., Veeranna, and Grant, P. 2000. Regulation of axonal neurofilament phosphorylation. In *Current Topics in Cellular Regulation, Vol 36*. ACADEMIC PRESS INC.

Pearson, G., Robinson, F., Gibson, T. B., Xu, B. E., Karandikar, M., Berman, K., and Cobb, M. H. 2001. Mitogen-activated protein (map) kinase pathways: Regulation and physiological functions. *Endocrine Reviews*, 22(2):153–183.

Pendry, J. B. 2000. Negative refraction makes a perfect lens. *Phys. Rev. Lett.*, 85(18):3966–3969.

Perrot, R., Berges, R., Bocquet, A., and Eyer, J. 2008. Review of the multiple aspects of neurofilament functions, and their possible contribution to neurodegeneration. *Molecular Neurobiology*, 38(1):27–65.

Perry, G., Nunomura, A., Raina, A. K., Aliev, G., Siedlak, S. L., Harris, P. L. R., Casadesus, G., Petersen, R. B., Bligh-Glover, W., Balraj, E., Petot, G. J., and Smith, M. A. 2003. A metabolic basis for alzheimer disease. *Neurochemical Research*, 28(10):1549–1552.

Petitpas, I., Grune, T., Bhattacharya, A. A., and Curry, S. 2001. Crystal structures of human serum albumin complexed with monounsaturated and polyunsaturated fatty acids. *Journal of Molecular Biology*, 314(5):955–960.

Bibliography

Pike, L. J. and Miller, J. M. 1998. Cholesterol depletion delocalizes phosphatidylinositol biphosphate and inhibits hormone-stimulated phosphatidylinositol turnover. *J Biol Chem*, 273:22298–22304.

Pörn, M. I. and Slotte, J. P. 1995. Localization of cholesterol in sphingomyelinase-treated fibroblasts. *Biochem. J.*, 308(1):269–274.

Qi, M. S., Liu, Y. Z., Freeman, M. R., and Solomn, K. R. 2009. Cholesterol-regulated stress fiber formation. *Journal of Cellular Biochemistry*, 106(6):1031–1040.

Redman, D. A., Brown, S., Sands, R. H., and Rand, S. C. 1991. Spin dynamics and electronic states of n-v centers in diamond by epr and four-wave-mixing spectroscopy. *Physical Review Letters*, 67:3420 – 3423.

Resjo, S., Oknianska, A., Zolnierowicz, S., Manganiello, V., and Degerman, E. 1999. Phosphorylation and activation of phosphodiesterase type 3b (pde3b) in adipocytes in response to serine/threonine phosphatase inhibitors: deactivation of pde3b in vitro by protein phosphatase type 2a. *Biochemical Journal*, 341:839–845.

Rigler, R. and Elson, E. S., e. 2001. *Fluorescence correlation spectroscopy. Theory and applications*, volume 65 of *Springer Series in Chemical Physics*. Springer-Verlag, Berlin, Heidelberg.

Rigler, R., Mets, ., Widengren, J., and Kask, P. 1993. Fluorescence correlation spectroscopy with high count rate and low background: analysis of translational diffusion. *Eur. Biophys. J.*, 22:169–175.

Rigler, R. and Widengren, J. 1990. Ultrasensitive detection of single molecules by fluorescence correlation spectroscopy. *BioScience, Lund University Press*, pages 180–183.

Ringemann, C. 2008. *Single molecule studies at the nanoscale: STED Fluorescence Fluctuation Spectroscopy in subdiffraction focal volumes*. PhD thesis, Georg-August University of Göttingen.

Rittweger, E., Han, K. Y., Irvine, S. E., Eggeling, C., and Hell, S. W. 2009. Sted microscopy reveals crystal colour centres with nanometric resolution. *Nature Photonics*, 3(3):144–147.

Ross, R. A., Spengler, B. A., and Biedler, J. L. 1983. Coordinate morphological and biochemical interconversion of human neuro-blastoma cells. *Journal of the National Cancer Institute*, 71(4):741–749.

Rust, M. J., Bates, M., and Zhuang, X. 2006. Sub-diffraction-limit imaging by stochastic optical reconstruction microscopy (storm). *Nat. Methods*, 3:793–796.

Saito, T., Shima, H., Osawa, Y., Nagao, M., Hemmings, B. A., Kishimoto, T., and Hisanaga, S. 1995. Neurofilament-associated protein phosphatase 2a - its possible role in preserving neurofilaments in filamentous states. *Biochemistry*, 34(22):7376–7384.

Salinovich, O. and Montelaro, R. C. 1986. Reversible staining and peptide-mapping of proteins transferred to nitrocellulose after separation by sodium dodecyl-sulfate polyacrylamide-gel electrophoresis. *Analytical Biochemistry*, 156(2):341–347.

Sasaki, T., Gotow, T., Shiozaki, M., Sakaue, F., Saito, T., Julien, J. P., Uchiyama, Y., and Hisanaga, S. I. 2006. Aggregate formation and phosphorylation of neurofilament-l pro22 charcot-marie-tooth disease mutants. *Human Molecular Genetics*, 15(6):943–952.

Saxton, M. 1994. Anomalous diffusion due to obstacles: a monte carlo study. *Biophys. J.*, 66:394–401.

Saxton, M. and Jacobson, K. 1997. Single particle tracking: Applications to membrane dynamics. *Annu. Rev. Biophys. Biomol. Struct.*, 26:373–399.

Schönle, A. and Hell, S. W. 1998. Heating by absorption in the focus of an objective lens. *Opt. Lett.*, 23(5):325–327.

Schütz, G., Kada, G., Pastushenko, V., and Schindler, H. 2000. Properties of lipid microdomains in a muscle cell membrane visualized by single molecule microscopy. *EMBO J.*, 19(5):892–901.

Schwille, P., J., K., and Webb, W. 1999. Fluorescence correlation spectroscopy with single-molecule sensitivity on cell and model membranes. *Cytometry*, 36:176–182.

Bibliography

Schwille, P., Meyer-Almes, F. J., and Rigler, R. 1997. Dual-color fluorescence cross-correlation spectroscopy for multicomponent diffusional analysis in solution. *Biophys. J.*, 72(4):1878–1886.

Sharma, M., Sharma, P., and Pant, H. C. 1999. Cdk-5-mediated neurofilament phosphorylation in shsy5y human neuroblastoma cells. *Journal of Neurochemistry*, 73(1):79–86.

Shera, E. B., Seitzinger, N. K., Davis, L. M., Keller, R. A., and Soper, S. A. 1990. Detection of single fluorescent molecules. *Chem. Phys. Lett.*, 174(6):553–557.

Sihag, R. K., Jaffe, H., Nixon, R. A., and Rong, X. H. 1999. Serine-23 is a major protein kinase a phosphorylation site on the amino-terminal head domain of the middle molecular mass subunit of neurofilament proteins. *Journal of Neurochemistry*, 72(2):491–499.

Simons, K. and Ikonen, E. 1997. Functional rafts in cell membranes. *Nature*, 387:569–572.

Simons, K. and Vaz, W. L. C. 2004. Model systems, lipid rafts, and cell membranes. *Annual Review of Biophysics and Biomolecular Structure*, 33:269–295.

Slawson, C. and Hart, G. W. 2003. Dynamic interplay between o-glcnac and o-phosphate: the sweet side of protein regulation. *Current Opinion in Structural Biology*, 13(5):631–636.

Small, D. M. 1984. Lateral chain packing in lipids and membranes. *Journal of Lipid Research*, 25(13):1490–1500.

Spasic, M. R., Callaerts, P., and Norga, K. K. 2009. Amp-activated protein kinase (ampk) molecular crossroad for metabolic control and survival of neurons 10.1177/1073858408327805. *Neuroscientist*, 15(4):309–316.

Spiegel, S. and Milstien, S. 2002. Sphingosine 1-phosphate, a key cell signaling molecule 10.1074/jbc.r200007200. *J. Biol. Chem.*, 277(29):25851–25854.

Sprague, B., Pego, R., Stavreva, D., and McNally, J. 2004. Analysis of binding reactions by fluorescence recovery after photobleaching. *Biophys J*, 86:3473–3495.

Sternberger, N. H., Sternberger, L. A., and Ulrich, J. 1985. Aberrant neurofilament phosphorylation in alzheimer-disease. *Proceedings of the National Academy of Sciences of the United States of America*, 82(12):4274–4276.

Stokes, G. 1852. On the refrangibility of light. *Phil. Trans.*, 142:463–562.

Strack, S., Westphal, R. S., Colbran, R. J., Ebner, F. F., and Wadzinski, B. E. 1997. Protein serine/threonine phosphatase 1 and 2a associate with and dephosphorylate neurofilaments. *Molecular Brain Research*, 49(1-2):15–28.

Sun, D. M., Leung, C. L., and Liem, R. K. H. 1996. Phosphorylation of the high molecular weight neurofilament protein (nf-h) by cdk5 and p35. *Journal of Biological Chemistry*, 271(24):14245–14251.

Synge, E. 1928. A suggested method for extending microscopic resolution into the ultramicroscopic region. *Philos. Mag.*, 6:356.

Tang, Q. and Edidin, M. 2003. Lowering the barriers to random walks on the cell surface. *Biophysical Journal*, 84(1):400–407.

Tapscott, S. J., Bennett, G. S., Toyama, Y., Kleinbart, F., and Holtzer, H. 1981. Intermediate filament proteins in the developing chick spinal-cord. *Developmental Biology*, 86(1):40–54.

Thewalt, J. and Bloom, M. 1992. Phosphatidylcholine: cholesterol phase diagrams. *Biophys J*, 63:1176–1181.

Tohyama, T., Lee, V. M. Y., Rorke, L. B., Marvin, M., McKay, R. D. G., and Trojanowski, J. Q. 1992. Nestin expression in embryonic human neuroepithelium and in human neuroepithelial tumor-cells. *Laboratory Investigation*, 66(3):303–313.

Troller, U., Zeidman, R., Svensson, K., and Larsson, C. 2001. A pkc beta isoform mediates phorbol ester-induced activation of erk1/2 and expression of neuronal differentiation genes in neuroblastoma cells. *Febs Letters*, 508(1):126–130.

Tsui-Pierchala, B. A., Encinas, M., Milbrandt, J., and Johnson, E. M. 2002. Lipid rafts in neuronal signaling and function. *Trends in Neurosciences*, 25(8):412–417.

Bibliography

Ulrich, J., Haugh, M., Anderton, B. H., Probst, A., Lautenschlager, C., and His, B. 1987. Alzheimer dementia and picks disease - neurofibrillary tangles and pick bodies are associated with identical phosphorylated neurofilament epitopes. *Acta Neuropathologica*, 73(3):240–246.

van Meer, G. and Liskamp, R. M. J. 2005. Brilliant lipids. *Nature Methods*, 2(1):14–15.

Veeranna, Amin, N. D., Ahn, N. G., Jaffe, H., Winters, C. A., Grant, P., and Pant, H. C. 1998. Mitogen-activated protein kinases (erk1,2) phosphorylate lys-ser-pro (ksp) repeats in neurofilament proteins nf-h and nf-m. *Journal of Neuroscience*, 18(11):4008–4021.

Veeranna, Shetty, K. T., Link, W. T., Jaffe, H., Wang, J., and Pant, H. C. 1995. Neuronal cyclin-dependent kinase-5 phosphorylation sites in neurofilament protein (nf-h) are dephosphorylated by protein phosphatase 2a. *J Neurochem*, 64(6):2681–90.

Venable, M. E., Webb-Froehlich, L. M., Sloan, E. F., and Thomley, J. E. 2006. Shift in sphingolipid metabolism leads to an accumulation of ceramide in senescence. *Mechanisms of Ageing and Development*, 127(5):473–480.

Wachsmuth, M., Waldeck, W., and Langowski, J. 2000. Anomalous diffusion of fluorescent probes inside living cell nuclei investigated by spatially-resolved fluorescence correlation spectroscopy. *J. Mol. Biol.*, 298:677–689.

Wagner, L. M. and Danks, M. K. 2009. New therapeutic targets for the treatment of high-risk neuroblastoma. *Journal of Cellular Biochemistry*, 107(1):46–57.

Wang, J. Z., Tung, Y. C., Wang, Y. P., Li, X. T., Iqbal, K., and Grundke-Iqbal, I. 2001. Hyperphosphorylation and accumulation of neurofilament proteins in alzheimer disease brain and in okadaic acid-treated sy5y cells. *Febs Letters*, 507(1):81–87.

Wang, T.-Y. and Silvius, J. R. 2000. Different sphingolipids show differential partitioning into sphingolipid/cholesterol-rich domains in lipid bilayers. *Biophys J*, 79:1478–1489.

Wawrezinieck, L., Rigneault, H., Marguet, D., and Lenne, P.-F. 2005. Fluorescence correlation spectroscopy: diffusion laws to probe the submicron cell membrane organization. *Biophys J*, 89:4029–4042.

Wells, L., Kreppel, L. K., Comer, F. I., Wadzinski, B. E., and Hart, G. W. 2004. O-glcnac transferase is in a functional complex with protein phosphatase 1 catalytic subunits. *J. Biol. Chem.*, 279(37):38466–38470.

Wenger, J., Conchonaud, F., Dintinger, J., Wawrezinieck, L., Ebbesen, T., Rigneault, H., Marguet, D., and Lenne, P.-F. 2007. Diffusion analysis within single nanometric apertures reveals the ultrafine cell membrane organization. *Biophys. J.*, 92(3):913–919.

Widengren, J., Rigler, R., and Mets, . 1994. Triplet-state monitoring by fluorescence correlation spectroscopy. *J. Fluoresc.*, 4:255–258.

Wildanger, D., Rittweger, E., Kastrup, L., and Hell, S. W. 2008. Sted microscopy with a supercontinuum laser source. *Opt. Express Optics Express*, 16(13):9614–9621.

Winder, W. W. and Hardie, D. G. 1996. Inactivation of acetyl-coa carboxylase and activation of amp-activated protein kinase in muscle during exercise. *Am J Physiol Endocrinol Metab*, 270(2):E299–304.

Xu, Z., Marszalek, J., Lee, M., Wong, P., Folmer, J., Crawford, T., Hsieh, S., Griffin, J., and Cleveland, D. 1996. Subunit composition of neurofilaments specifies axonal diameter. *J. Cell Biol.*, 133(5):1061–1069.

Yechiel, E. and Edidin, M. 1987. Micrometer-scale domains in fibroblast plasma membranes. *J. Cell Biol.*, 105(2):755–760.

Yethiraj, A. and Weisshaar, J. 2007. Why are lipid rafts not observed in vivo? *Biophys J BioFAST*, 93:3113–3119.

Yuan, A. D., Rao, M. V., Sasaki, T., Chen, Y. X., Kumar, A., Veeranna, Liem, R. K. H., Eyer, J., Peterson, A. C., Julien, J. P., and Nixon, R. A. 2006. alpha-internexin is structurally and functionally associated with the neurofilament triplet proteins in the mature cns. *Journal of Neuroscience*, 26(39):10006–10019.

Zacharias, D., Violin, J., Newton, A., and Tsien, R. 2002. Partitioning of lipid-modified monomeric gfps into membrane microdomains of live cells. *Science*, 296:913–916.

Zander, C., Enderlein, J., and Keller (eds.), R. A. 2002. *Single-molecule detection in solution*. Wiley-VCH, Berlin, Germany, 1st edition.

Bibliography

Zhang, N., Gao, G., Bu, X. N., Han, S., Fang, L., and Li, J. F. 2007. Neuron-specific phosphorylation of c-jun n-terminal kinase increased in the brain of hypoxic preconditioned mice. *Neuroscience Letters*, 423(3):219–224.

Zheng, Y. L., Li, B. S., Veeranna, and Pant, H. C. 2003. Phosphorylation of the head domain of neurofilament protein (nf-m) - a factor regulating topographic phosphorylation of nf-m tail domain ksp sites in neurons. *Journal of Biological Chemistry*, 278(26):24026–24032.

Zidovetzki, R. and Levitan, I. 2007. Use of cyclodextrins to manipulate plasma membrane cholesterol content: Evidence, misconceptions and control strategies. *Biochimica et Biophysica Acta (BBA) - Biomembranes*, 1768(6):1311 – 1324.

7 Supplementary

Antibody	Host Ig isotype	Company
[V9] Anti-Vimentin	mouse IgG	Sigma V6389
Anti-Doublecortin	rabbit IgG	Abcam ab18723
[SMI-33] Anti-Neurofilament H&M non-phosphorylated epitope	mouse IgM	Covance SMI-33R
[SMI-34] Anti-Neurofilament H&M phosphorylated epitope	mouse IgG	Covance SMI-34R
[SMI-36] Anti-Neurofilament H phosphorylated epitope	mouse IgG	Abcam ab24572
[NL6] Anti-Neurofilament M O-GlcNAcylated epitope	mouse IgG	Sigma
Anti-α-Tubulin-biotin-XX	mouse IgG	Invitrogen A21371
[TUB 2.1] Anti-β-Tubulin	mouse IgG	Sigma T4026
Anti-GFAP-biotin	mouse IgG	Dianova DLN-08735
Anti-mouse	sheep IgG	Dianova 515-005-003
Anti-rabbit	goat IgG	Dianova 111-005-003

Table 7.1: Antibodies used for immunofluorescence. All antibodies were used in a working concentration of $1 - 5\mu g/\mu l$

Dye	MW [g/mol]	λ_{abs} [nm]	ϵ_{max} [M^{-1} cm^{-1}]	CF$_{260}$	CF$_{280}$
Atto590-NHS	788	594	1.1 x 10^5	0.42	0.44
Atto594-NHS	1389	601	1.2 x 10^5	0.26	0.51
Atto647N-NHS	843	644	1.5 x 10^5	0.06	0.05
Atto647N-Maleimide	868	644	1.5 x 10^5	0.06	0.05

Table 7.2: Optical properties of the used NHS- and Maleimide-dyes. To calculate the degree of labeling (DOL) the following parameters need to be known: the maximum absorption wavelength (λ_{abs}), the extinction coefficient of the free dye at the absorption maximum (ϵ_{max}), the correction factors at 260 nm and 280 nm (CF$_{260}$ and CF$_{280}$), respectively. The extinction coefficient for an IgG molecule is 210000 M^{-1}cm^{-1}, for the neurofilament heavy subunit (NFH) 28420 M^{-1}cm^{-1} and for the light subunit NFL 36790 M^{-1}cm^{-1}.

Supplementary

Abbreviation	Systematic name	MW [Da]	Synthesized by
PE-Atto647N	N-(Atto647N)-(1,2-dihexadecanoyl-*sn*-glycero-3-phosphoethanolamine)	1492.06	Atto-Tec
PE-1-Atto647N	1-hexadecanoyl-2-Atto647N-*sn*-	1182.35	Vladimir Belov
DOPE-Atto647N	N-(Atto647N)-(1,2-dioctadecenoyl-*sn*-glycero-3-phosphoethanolamine)	1490.06	Atto-Tec
CPE-Atto647N	N-dodecanoyl-D-*erythro*-sphingosyl-[N'-(Atto647N)-phosphoethanolamine] (C_{17} base) = Atto647N-ceramide-phosphoethanolamine glycero-3-phosphoethanolamine)	1238.85	Atto-Tec
SM-Atto647N	N-(Atto647N)-sphingosyl-phosphocholine = Atto647N-sphingomyelin	1477.15	Atto-Tec
SM-BODIPY	BODIPY FL C_5-sphingosyl-phosphocholine = BODIPY FL C_5-sphingomyelin	766.75	Invitrogen
SM-NBD	NBD C_6-sphingosyl-phosphocholine = NBD C_6-sphingomyelin	740.88	Invitrogen
SM-Atto532	Atto532-sphingosyl-phosphocholine = Atto532-sphingomyelin	1477.15	Atto-Tec
GM1-Atto647N	Atto647N-monosialotetrahexosyl-ganglioside	1986.79	Günter Schwarzmann
GM1-BODIPY	BODIPY-FL C_5-monosialotetrahexosyl-ganglioside	1582.50	Invitrogen
GM1-NBD	NBD-C_6-monosialotetrahexosyl-ganglioside	1572.78	Günter Schwarzmann
GM1-#-Atto647N	Atto647N-monosialotetrahexosyl-ganglioside	2158	Svetlana Polyakova
GM1-##-Atto647N	Atto647N-monosialotetrahexosyl-ganglioside (C_{18} and C_{20})	2187 2215	Svetlana Polyakova
Cer-Atto647N	Atto647N-D-*erythro*-sphingosine (Ceramide)	1016.53	Atto-Tec

Table 7.3: Lipids used for BSA complexes

NFH primary sequence

Homo sapiens	1	MMSFGADALLGAPFAPLHGGGSLHYALARKGGAGGTRSAAGSSSGFHSWTRTSVSSVSASPSRFRGAGAASSTDSLDTL
Mus musculus	1	MMSFGSADALLGAPFAPLHGGGSLHYSLSRKAGPGGTRSAAGSSSGFHSWARTSVSSVSASPSRFRGA--ASSTDSLDTL
Bos taurus	1	-MSFSGADALLGA-FAPLHGGGSLHYALARKGGAGGARSAAGSSSGFHSWARTSVSSVSASPSRFRGAGTTSSTDSLDTL
Homo sapiens	81	SNGPEGCMVAVATSRSEKEQLQALNDRFAGYIDKVRQLEAHNRSLEGEAAALRQQQAGRSAMGELYEREVREMRGAVLRL
Mus musculus	79	SNGPEGCVVAAVAARSEKEQLQALNDRFAGYIDKVRQLEAHNRSLEGEAAALRQQQAGRAAMGELYEREVREMRGAVLRL
Bos taurus	79	SNGPEGCVVV---ARSEKEQLQVLNDRFAGYIDKVRQLEAHNRSLEGEAAALRQQQAGRAAMGELYEREVREMRGAVLRL
Homo sapiens	161	GAARGQLRLEQEHLLEDIAHVRQRLDDEARQREEAEAAARALRFAQEAEAARVDLQKKAQALQEECGYLRRHHQEEVGE
Mus musculus	159	GAARGQLRLEQEHLLEDIAHVRQRLDEEARQREEAEAAARALRFAQEAEAARVELQKKAQALQEECGYLRRHHQEEVGE
Bos taurus	156	GAARGQLRLEQEHLLEDIAHVRQRLDDEARQREEAAATRALARFAQEAEAARVELQKKAQALQEECGYLRRHHQEEVGE
Homo sapiens	241	LLGQIQGSGAAQAQMQAETRDALKCDVTSALREIRAQLEGHAVQSTLQSEEWFRVRLDRLSEAAKVNTDAMRSAQEEITE
Mus musculus	239	LLGQIQGCGAAQAQAQAEARDALKCDVTSALREIRAQLEGHAVQSTLQSEEWFRVRLDRLSEAAKVNTDAMRSAQEEITE
Bos taurus	236	LLGQIQSSSAAQTQ--AEARDALKCDVTSALREIRAQLEGHTVQSTLQSEEWFRVRLDRLSEAAKVNTDAMRSAQEEISE
Homo sapiens	321	YRRQLQARTTELEALKSTKDSLERQRSELEDRHQADIASYQEAIQQLDAELRNTKWEMAAQLREYQDLLNVKMALDIEIA
Mus musculus	319	YRRQLQARTTELEALKSTKESLERQRSELEDRHQADIASYQDAIQQLDSELRNTKWEMAAQLREYQDLLNVKMALDIEIA
Bos taurus	314	YRRQLQARTTELETLKSTKDSLERQRSELEDRHQADIASYQEAIQQLDAELRNTKWEMAAQLREYQDLLNVKMALDIEIA
Homo sapiens	401	AYRKLLEGEECRIGFGPIPFSLPEGLPKIPSVSTHIKVKSEEKIKVVEKSEKETVIVEEQTEETQVTEEVTEEEEKEAKE
Mus musculus	399	AYRKLLEGEECRIGFGPSPFSLTEGLPKIPSISTHIKVKSEEMIKVVEKSEKETVIVEGQTEEIRVTEGVTEEEDKEAQG
Bos taurus	394	AYRKLLEGEECRIGFGPSPFLLPEGLPKIPSVSTHIKVKSEEKIKMVEKSEKETVIVEEQTEEVQVTEEVTEEEEKEAKE
Homo sapiens	481	EEGKEEEGGEEEEAEGGEEETKSPPAEEAASPEK-------EAKSPVKEEAKSPAEAKSPEKE----------EAKSPAE
Mus musculus	479	QEGEEAEEGEEKEEEEGAAAT-SPPAEEAASPEK-------ETKSRVKEEAKSPGEAKSPGEAKSPAEAKSPGEAKSPGE
Bos taurus	474	EEGEAVKSPPAEEAASPEKEEAKSPA-EAKSPEKAKSPVKEEAKSPVKEEAKSPAEVKSPEK-----------AKSPVE
Homo sapiens	544	VKSPEKAKSPAK----EEAKSPPEAKSPEK----------------EEAKSPAEVKSPEKAKSPAKE----EAKSPAEAK
Mus musculus	551	AKSPGEAKSPAEPKSPAEPKSPAEAKSPAEPKSPATVKSPGEAKSPSEAKSPAEAKSPAEAKSPAEAKSPAEAKSPAEAK
Bos taurus	541	AKSPEKAKSPVK-----EEAKSPEKAKSPEK----------------EEAKSPEKAKSPEKAKSPVKE----EAKSPVEAK
Homo sapiens	600	SPEKAKSPVK---------EEAKSPAEAKSPVKE----EAKSPA--------EVKSPEKAKSPTK----EEAKSPEKAK
Mus musculus	631	SPAEAKSPATVKSPGEAKSPSEAKSPAEAKSPAEAKSPA---------EVKSPGEAKSPAEPKSPAEAKSPAEVK
Bos taurus	597	SPEKAKSPVK---------EEAKSPEKAKSPVKE----EAKSPEKAKSPVKEEAKSPEKAKSPVK----EEAKSPEKAK
Homo sapiens	654	SPEKE---------EAKSPEKAKSPVKA---------EAKSPEKAKSPVKAEAKSPEKAKSPVKEEAKSPEKAKSPVK
Mus musculus	703	SPAEAKSPAEVKSPGEAKSPAAVKSPAEAKSPAAVKSPGEAKSPGEAKSPA--EAKSPAEAKSPI--EVKSPEKAKTPVK
Bos taurus	659	SPVKE---------EAKSPEKAKSPVKE----------EAKSPEKAKSPV--EAKSPEKAKSPAKEEAKSPEKAKSPVK
Homo sapiens	714	E------EAKSPEKAKSPVKEEAKTPEKAKSPVKEEAKSPEKAKSPEKAKTL---DVKSPEA-KTPAKEEARSPAD-KFP
Mus musculus	779	EGAKSPAEAKSPEKAKSPVKEDIKPPAEAKSPEKAKSPVKEGAKPPEKAKPL---DVKSPEA-QTPVQEEAKHPTDIRPP
Bos taurus	717	E------EAKSPEKAKSPVKEAKSPEKAKSPAKEEAKSPEKAKSPEKPKSPVKEEAKSPEKPKSPVKEEAKS------P
Homo sapiens	783	EKAKSPVKEEVKSPEKAKSPLKEDAKAPEKEIPKKEEVKSPVKEEEKPQEVKVKEPPKKAEEEKAPATPKTEEKKDSKKE
Mus musculus	855	EQVKSPAKE------KAKSPEKEEAKTSEKVAPKKEEVKSPVKE-----EVKAKEPPKKVEEEKTLPTPKT-EAKESKKD
Bos taurus	785	EKPKSPVKEEAKSPEKAKSPVKDEAKAPGKEVLKKEEAKSPVKEEEKPQEVRAKEPPKKAEEEKAPAIPKAEEKKDSKKD
Homo sapiens	863	EAPKKEAPKPKVEEKKEPAVEKPKESKVEAKKEEAEDKKKVPTPEKEAPAKVEVKEDAKPKEKTEVAKKEPDDAKAKEPS
Mus musculus	923	EAPK-EAPKPKVEEKKETPTEKPKDSTAEAKKEEAGEKKKAVASEEETPAKLGVKEEAKPKEKTETTKTEAEDTKAKEPS
Bos taurus	865	EVPKKEAPKP--EEKKEPAVEKPKEPKAEAKKEEAEDKKKATTPEKEAPAKG--KEEAKPKEKTEVAKKEPDDAKPKEPS
Homo sapiens	943	KPAE----KKE-----AAPEKKDTKEEK---AKKPEEKPKTEAKAKEDDKTLSKEPSKPKAEKAEKSSSTDQKDSKPPEK
Mus musculus	1002	KPTETEKPKKE--EMPAAPEKKDTKEEKTTESRKPEEKPKMEAKVKEDDKSLSKEPSKPKTEKAEKSSSTDQKESQPPEK
Bos taurus	941	KAAE----KPKKEETPAAPEKKDAKEGKAAEAKKPEEKPKTEAQAKEP----GKEASTPGKAKAEKSSSTDQKDSRPAEK
Homo sapiens	1011	ATEDKAAKGK- 1020
Mus musculus	1080	TTEDKATKGEK 1090
Bos taurus	1013	AAEDKATKGEK 1023

Figure 7.1: Primary sequences of human, mouse and bovine NFH. The sequences were blasted using the NCBI protein blast algorithm. They can be found under the accordant NCBI reference numbers. Homo sapiens: NP_066554, Mus musculus: NP_035034, Bos taurus: XP_870725. The bovine NFH sequence is predicted.

Supplementary

8 List of Publications

8.1 Publications Related to Thesis

Wildanger, D., **R. Medda**, L. Kastrup and S.W. Hell. *A compact STED microscope providing 3D nanoscale resolution.*, Journal of Microscopy, **2009**

Eggeling*, C., C. Ringemann*, **R. Medda**, G. Schwarzmann, K. Sandhoff, S. Polyakova, V.N. Belov, B. Hein, C. von Middendorff, A. Schönle and S.W. Hell, * equally contributed. *Direct observation of the nanoscale dynamics of membrane lipids in a living cell.*, Nature, 457:1159-1163, **2009**

Meyer, L., D. Wildanger, **R. Medda**, A. Punge, S.O. Rizzoli, G. Donnert and S.W. Hell. *Dual-color STED microscopy at 30nm focal-plane resolution.*, Small, 4(8):1095-1100, **2008**

Donnert, G., J. Keller, **R. Medda**, M.A. Andrei, S.O. Rizzoli, R. Lührmann, R. Jahn, C. Eggeling and S.W. Hell. *Macromolecular-scale resolution in biological fluorescence microscopy.*, Proc. Natl. Acad. Sci. USA, 103(31):11440-11445, **2006**

awarded with Cozzarelli prize

8.2 Publications in Cooperation (Selection)

Testa, I., A. Schönle, C. von Middendorff, C. Geisler, **R. Medda**, C.A. Wurm, A.C. Stiel, S. Jakobs, M. Bossi, C. Eggeling, S.W. Hell and A. Egner. *Nanoscale separation of molecular species based on their rotational mobility.*, Opt. Express, 16(25):21093-104, **2008**

Fölling, J., M. Bossi, H. Bock, **R. Medda**, C.A. Wurm, B. Hein, S. Jakobs, C. Eggeling and S.W. Hell. *Fluorescence nanoscopy by ground-state depletion and single-molecule return.*, Nat. Methods, 5(11):943-5, **2008**

M. Bossi, J. Fölling, V.N. Belov, V.P. Boyarskiy, **R. Medda**, A. Egner, C. Eggeling, A. Schönle and S.W. Hell. *Multicolor Far-Field Fluorescence Nanoscopy through Isolated Detection of Distinct Molecular Species.* Nano Lett., 8(8):2463-8, **2008**

Boyarskiy, V.P., V.N. Belov, **R. Medda**, B. Hein, M. Bossi and S.W. Hell. *Photostable, amino reactive and water-soluble fluorescent labels based on sulfonated rhodamine with a rigidized xanthene fragment.*, Chemistry, 14(6):1784-92, **2008**

Egner, A., C. Geisler, C. von Middendorff, H. Bock, D. Wenzel, **R. Medda**, M. Andresen, A.C. Stiel, S. Jakobs, C. Eggeling, A. Schönle and S.W. Hell. *Fluorescent nanoscopy in whole cells by asynchronous localization of photoswitching emitters.*, Biophys J., 93(9):3285-90, **2007**

Willig, K.I., B. Harke, **R. Medda** and S.W. Hell. *STED microscopy with continuous wave beams.*, Nat. Methods, 4(11):915-8, **2007**

Fölling, J., V.N. Belov, R. Kunetsky, **R. Medda**, A. Schönle, A. Egner, C. Eggeling, M. Bossi and S.W. Hell. *Photochromic rhodamines provide nanoscopy with optical sectioning.*, Angew. Chem. Int. Ed. Engl., 46(33):6266-70, **2007**

9 Contribution to Conferences

R. Medda, L. Kastrup, D. Wildanger and S.W. Hell. "Application of STED Microscopy for Neurofilament studies in re-differentiated human Neuroblastoma.", Focus on Microscopy **2009**, Krakow
oral presentation

R. Medda, D. Wildanger, L. Kastrup and S.W. Hell. "STED microscopy as a tool for Neurofilament studies in redifferentiated Neuroblastoma.", Gordon Research Conference, Intermediate Filaments **2008**, Oxford
poster presentation, awarded with best poster prize

R. Medda, J. Bewersdorf and S.W. Hell. "4Pi microscopy with Quantum Dots.", Deutsche Gesellschaft für Zellbiologie **2007**, Frankfurt am Main
poster presentation

R. Medda, J. Bewersdorf and S.W. Hell. "4Pi-microscopy Type C with Quantum Dots", Jackson Lab Bar Harbor **2006**
invited oral presentation

Contribution to Conferences

10 Acknowledgment

First I would like to thank Prof. Stefan Hell for giving me the opportunity to work on these projects, his motivation and interest in this work and for providing outstanding lab environment.

I also thank Prof. Klaus-Armin Nave for his interest in this work and the willingness to be the main referee, and furthermore, for his encouragement and tips, especially in the neurofilament project.

Furthermore, I would like to take this opportunity to thank all those people who have contributed to the success of this work:

- Christian Eggeling and Christian Ringemann for the great collaboration in the lipid project
- Claas von Middendorff for the analysis of the lipid dynamics
- Vladimir Belov and Svetlana Polyakova for synthesizing many dyes and lipids
- Günther Schwarzmann for synthesizing the labeled gangliosides
- Thorsten Lang for teaching me how to generate liposomes and membrane sheets
- Heinz-Jürgen Dehne and Mary Osborn for introducing me to the world of cell culture
- The group of Prof. Nave, and in particular Celia Kassmann, for fruitful discussions and proposals
- Kyu-Young Han, Vladimir Belov, Gyuzel Mitronova, Heiko Schill and Christian Eggeling for the ongoing collaboration in the nanodiamond project

Acknowledgment

- Lars Kastrup and Dominik Wildanger for providing me their supercontinuum STED setup

- Jan Keller and Veronika Müller for the help with Latex

- Christian Wurm, Arnold Giske, Veronika Müller, Jaydev Jethwa, Katrin Willig, Lars Kastrup, Eva Rittweger, Marcel Lauterbach, Jan Keller and Brian Rankin for proofreading parts of the manuscript

- Sarah Aschemann and Josephine Stadler for selfless aid and excellent organization

- Marco Roose for solving my numerous computer problems

- Ellen Rothermel and Tanja Gilat for excellent and unimpeachable assistance

- Maria Sermond for a constant supply of clean glass ware, pipet tips, liquid nitrogen and happiness

- All people, who made the time unforgettable here at MPI, with chamber music, barbecues, horse back riding and holidays

- My present and former office mates Claudia Geisler, Ilaria Testa, Tobias Müller and Astrid Schauss for the nice atmosphere

- My family, especially my parents and Arnold, for their constant support and encouragement

Die VDM Verlagsservicegesellschaft sucht für wissenschaftliche Verlage abgeschlossene und herausragende

Dissertationen, Habilitationen, Diplomarbeiten, Master Theses, Magisterarbeiten usw.

für die kostenlose Publikation als Fachbuch.

Sie verfügen über eine Arbeit, die hohen inhaltlichen und formalen Ansprüchen genügt, und haben Interesse an einer honorarvergüteten Publikation?

Dann senden Sie bitte erste Informationen über sich und Ihre Arbeit per Email an *info@vdm-vsg.de*.

Sie erhalten kurzfristig unser Feedback!

VDM Verlagsservicegesellschaft mbH
Dudweiler Landstr. 99
D - 66123 Saarbrücken

Telefon +49 681 3720 174
Fax +49 681 3720 1749

www.vdm-vsg.de

Die VDM Verlagsservicegesellschaft mbH vertritt

Printed by Books on Demand GmbH, Norderstedt / Germany